Basics of Reservoir Engineering

Basics of Reservoir Engineering

Editor

Sanjay Walia

scitus
academics

Basics of Reservoir Engineering

Edited by **Sanjay Walia**

Printed in 2017

ISBN: 978-1-68117-336-8

Library of Congress Control Number: 2015939248

© 2016 by

SCITUS Academics LLC,
616, Corporate Way, Suite 2, 4766,
Valley Cottage, NY 10989

www.scitusacademics.com

Contents

Preface vii

Chapter 1   Prediction of Molar Volumes of the Sudanese Reservoir Fluids 1

A. A. Rabah and S. A. Mohamed

Chapter 2   Field Scale Simulation Study of Miscible Water Alternating CO_2 Injection Process in Fractured Reservoirs...27

Mohammad Afkhami Karaei, Ali Ahmadi, Hooman Fallah, Shahrokh Bahrami Kashkooli, and Jahangir Talebi Bahmanbeglo

Chapter 3   Fully Compositional and Thermal Reservoir Simulation 63

Rustem Zaydullin, Denis V. Voskov, Scott C. James, Heath Henley, and Angelo Lucia

Chapter 4   Methanol treatment in Gas Condensate Reservoirs: A Modeling and Experimental Study 113

A. Asgari, M. Dianatirad, M. Ranjbaran, A.R. Sadeghi,and M.R. Rahimpour

Chapter 5   GeoSys.Chem: Estimate of Reservoir Fluid Characteristics as First Step in Geochemical Modeling of Geothermal Systems 157

Mahendra P. Verma

Chapter 6   A 3-D Water/Rock Chemical Interaction Model for Prediction of HDR/HWR Geothermal Reservoir Performance 193

Zhenzi Jing, Kimio Watanabe, onathan Willis-Richards, and Toshiyuki Hashida

Chapter 7   An Automated Approach for an Optimised Least Cost Solution of Reinforced Concrete Reservoirs Using Site Parameters.................257

A. Stanton and A.A. Javadi

Chapter 8   **Natural Gas Treating by Selective Adsorption: Material Science and Chemical Engineering Interplay** 273

Marco Tagliabue, David Farrusseng, Susana Valencia,
Sonia Aguado, Ugo Ravonb Caterina Rizzo, Avelino Corma,
and Claude Mirodatos

Citations...309
Index..313

Preface

Reservoir engineering is a branch of petroleum engineering that applies scientific principles to the drainage problems arising during the development and production of oil and gas reservoirs so as to obtain a high economic recovery. The working tools of the reservoir engineer are subsurface geology, applied mathematics, and the basic laws of physics and chemistry governing the behavior of liquid and vapor phases of crude oil, natural gas, and water in reservoir rock. Of particular interest to reservoir engineers is generating accurate reserves estimates for use in financial reporting to the SEC and other regulatory bodies. Other job responsibilities include numerical reservoir modeling, production forecasting, well testing, well drilling and workover planning, economic modeling, and PVT analysis of reservoir fluids.

Editor

Chapter 1

Prediction of Molar Volumes of the Sudanese Reservoir Fluids

A. A. Rabah[1] and S. A. Mohamed[2]

[1]Department of Chemical Engineering, University of Khartoum, Khartoum, Sudan

[2]Department of Petroleum Transportation & Refining Engineering, College of Petroleum Engineering and Technology, Sudan University of Science and Technology, Khartoum, Sudan

ABSTRACT

This paper provided important experimental PVT data of the Sudanese reservoir fluids. It includes composition analysis of 11 mixtures and about 148 PVT data points of constant mass expansion (CME) tests at pressures below the bubble point. The datasets are compared with eight equations of state (EOS), namely, Peng Robinson (PR), Soave-Redlich-Kwong (SRK), Lawal-Lake-Silberberg (LLS), Adachi-Lu-Sugie (ALS), Schmidt-Wenzel (SW), Patel-Teja (PT), Modified-Nasrifar-Moshfeghian (MNM), and Harmens-Knapp (HK). The results of comparison reveals that, with the exception of PR and ALS EOSs, all other EOSs yield consistently a higher average absolute percent deviation (AAPD) in the prediction of molar volume; it exceeds 20% by all mixtures. The grand average AAPD of all mixtures is 17 and 16 for PR and ALS, respectively. ALS is selected to represents the mixtures. It is modified by replacing the coefficient (Ωb_1) of the parameter (b_1) in the dominator of repulsive term by that of PR. This procedure enhanced the accuracy of ALS by 30 to 90% for individual mixtures and the grand average AAPD is significantly reduced from 16 to about 7.

INTRODUCTION

In the absence of the experimental PVT study, properties such as isothermal compressibility factor, z-factor, and formation volume factor, are determined from empirically derived correlations or equations of state (EOSs). The correlations are basically developed for crude from certain geographical region with certain hydrocarbon and nonhydrocarbon contents and API. Hence such correlations may not be valid to crude oils of geographical regions other than those for which these correlations have been developed. Although EOSs are generalized correlations, their validity to different crudes varies.

Adepoju [1] has made extensive study on Texas oil and found that Peng Robison (PR) [2], and Soave-Redlich-Kwong (SRK) [3] give a higher average absolute percent deviation (AAPD) in the prediction of the total volume of reservoir fluids. He obtained a good result when PR and SRK are modified by replacing the repulsion and attraction terms by that of Lawal-Lake-Silberberg (LLS) EOS [4]. Akberzadeh et al. [5] have investigated the Modified-Nasrifar-Moshfeghian (MNM) EOSs, PR, and SRK for Western Canadian heavy oils. They have shown that MNM without any volume correction predicted the densities with accuracy similar to SRK EOS with volume correction.

Jensen [6] found that Adachi-Lu-Sugie (ALS) EOS [7] is the most accurate for prediction of the phase behavior of well-defined hydrocarbon mixtures with and without a considerable content of CO_2 or N_2. The ALS EOS seems to be well suited for calculation of the phase equilibrium of reservoir fluids but often proves to give inaccurate predictions of the densities of hydrocarbon mixtures [8]. By incorporating the volume translation principle of Peneloux et al. [9], ALS equation was found to give good results for hydrocarbon mixtures with and without a considerable content of CO_2 or N_2 [10].

Pedersen et al. [10] developed a characterization procedure for SRK coupled with the volume correction term of Peneloux et al. [9]. This procedure does not need experimental data and generally gives good prediction of saturation points and vapor-liquid equilibrium. However, the model frequently calculates a too large liquid precipitation for gas condensate when simulating constant composition expansion experiments. In addition, prediction of liquid density is sometimes inaccurate, they added.

Almehaideb et al. [11] tested crude and gas samples from 17 UAE reservoirs and found that PR EOS predicted the density and bubble point pressure of UAE petroleum reservoir with an error of 9.28%. Yu and Chen [12] have evaluated PR and PT for binary and tertiary nonpolar and polar mixtures. For binary mixtures, the grand average AAPD for PR and RT is 9.69 and 10.39, respectively. For tertiary mixtures AAPD of PR and PT is 17.20 and 17.04, respectively. Kumer [13] has used 3100 data points of reservoir fluids mainly sweet and sour dry gases from various sources to evaluate the compressibility factor using eight EOSs, namely LLS, VDW, PR, RK, SRK, SW, PT, and Trebble-Bishnoi (TB). He concluded that LLS is superior to other EOSs in the prediction of z-factor.

Sudanese crude oil has come to surface on a commercial scale in the mid of 1990s, and there is little known in open literature about its PVT properties. Therefore the purpose of this work is to select an EOS that best represents the PVT data of the Sudanese reservoir fluids. The candidate EOSs includes PR, SRK, LLS, ALS, SW, PT, MNM, and HK.

PVT STUDY

Experimental PVT data were supplied by the Ministry of Energy and Mining, Sudan, for a number of wells representing different reservoirs. The data include compositional analysis of single carbon numbers of up to Eicosanes plus ($C20^+$) and Hexatriacontanes plus ($C36^+$) and PVT of CME of bottomhole samples. Table 1 shows the composition analysis of eleven reservoir fluids lumped up to $C7^+$. Samples 1 and 2 are taken from [14], samples 3 to 11 are presented for the first time. The data also include bubble point pressure, reservoir temperature and molecular weight, and specific gravity of $C7^+$. In some PVT reports, the molecular weight of C7$^+$ is not available; under such condition it is obtained using the material balance as

$$M_{C_n+} = \frac{M_a - \sum_1^{C+-1} z_i M_i}{z_{C_n+}},$$

(1)

where M_i and M_a are the component molecular weight and the apparent molecular weight, respectively, and z is the mole fraction. The plus fraction specific gravity is calculated as

$$\gamma_{C_n+} = \frac{1.008 M_{C_n+}}{42.43 + M_{C_n+}}.$$

(2)

The experimental error in pressure is $\pm$ 5 psi, temperature is $\pm$ 0.5°F, and cell volume is $\pm$ 0.3 cc as reported in PVT studies.

Table 1: Mixtures composition analysis and reservoir conditions

Comp.	1	2	3	4	5	6	7	8	9	10	11
N_2	0.350	0.590	0.250	0.227	0.210	1.124	0.330	0.424	0.434	0.092	0.077
CO_2	0.943	0.576	8.700	6.944	4.520	5.428	5.040	0.258	0.311	0.398	0.675
C1	21.048	41.840	1.750	1.445	1.310	0.926	1.330	25.373	12.893	14.666	29.035
C2	4.440	11.471	0.160	0.388	0.060	0.104	0.140	2.317	2.508	1.771	10.001
C3	4.525	7.631	0.110	0.664	0.050	0.444	0.110	0.700	1.176	0.145	7.747
iC4	3.173	1.384	0.020	0.252	0.020	0.195	0.030	1.524	2.180	0.43	1.229
nC4	3.287	1.885	0.040	0.392	0.030	0.199	0.060	0.239	0.234	0.03	4.102
iC5	2.704	1.320	0.010	0.306	0.010	0.169	0.020	2.550	1.497	0.441	1.177
nC5	2.233	1.687	0.020	0.381	0.020	0.397	0.030	0.119	0.257	0.06	2.249
C6	3.572	1.903	0.020	0.970	0.030	0.761	0.050	2.623	3.169	1.244	2.736
C7⁺	53.725	29.713	88.920	88.031	93.740	90.253	92.860	63.873	75.341	80.723	40.972
Total	100.00	100.00	100.00	100.00	100.00	100.00	100.00	100.00	100.00	100.00	100.00
MWC7+	180.6	204.4	336.50	220.59	320.93	232.61	310.70	151.80	140.82	143.22	275.3
gC7+	0.816	0.835	0.895	0.845	0.896	0.852	0.896	0.788	0.775	0.778	0.843
TR (°F)	172	223	234.0	236.0	231.0	235.0	247.0	179.3	165.0	165	244.0
PB (Psig)	1804	3366	295.0	344.0	172.0	276.0	196.0	2362.0	1500.0	1571	1970.3
γsat	0.704	0.588	0.806	0.784	0.806	0.758	0.805	0.869	0.785	0.8021	0.672

FLASH ALGORITHM

Although flash calculation procedure is well documented, Figure 1 shows the Flash algorithm used in this work. The input data include the reservoir pressure (P), bubble point pressure (P_b), reservoir temperature (Z_i), reservoir fluid composition (ω), and the density and the molecular weight of C7+. For pure hydrocarbon and nonhydrocarbon, the critical properties (T_c, P_c, z_c), the molecular weight (MW), and the acentric factor (ω) are obtained from the generalized properties tables [15]. For C7+ the critical properties and the acentric factor are estimated for a given molecular weight and a specific gravity from Lawal-Tododo-Heinze [16] correlations. Binary interaction parameters (BIPs) for hydrocarbon-hydrocarbon, nonhydrocarbon-hydrocarbon, and nonhydrocarbon-nonhydrocarbon systems are taken as zero because all samples contains small amount of CO_2 and N_2. The vapor and liquid molar volumes are calculated using the flash algorithm. Mathematically, the two-phase flash calculation is the solution of Rachford-Rice equation that satisfies the equal fugacity constraint $\Sigma(f_i^L / f_i^v - 1)^2 < \varepsilon$ [17]. Newton-Raphson iteration scheme was employed. The calculation is initiated with k-value obtained using Wilson correlation. If a convergence is not obtained -value is modified as $K_{i,n+1} = K_{i,n} f^L_{i,n} / f_{i,n}^v$ [18]. The flash program is executed for eight EOSs. These are PR, SRK, LLS, ALS, SW, PT, MNM, and HK EOSs.

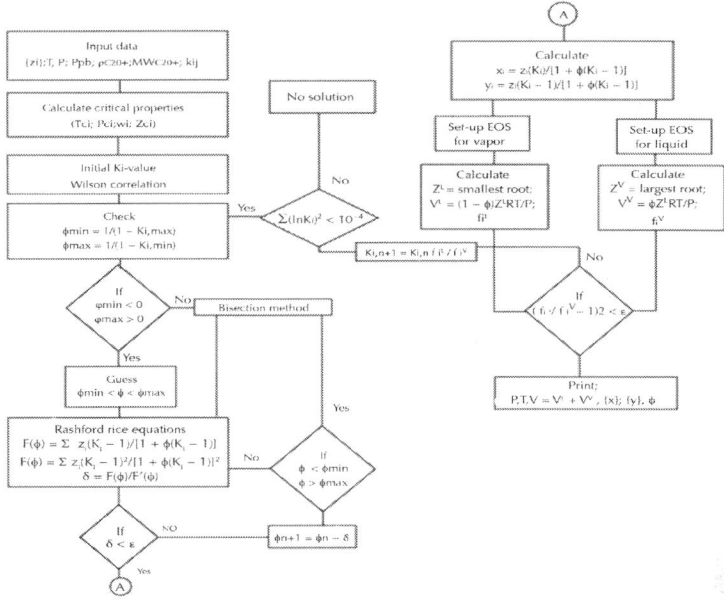

Figure 1: Flash algorithm.

RESULTS AND DISCUSSIONS

To compare EOSs to experimental data, numerous quality measurements based on statistical error analysis are computed. These include the percent deviation (PD), the average absolute percent deviation (AAPD), the minimum absolute percent deviation (APD_{min}), the maximum absolute percent deviation (APD_{max}), and the grand average AAPD. The percent deviation is defined as

$$PD_i = \frac{V_{exp} - V_{cal}}{V_{exp}} \times 100\%,$$

(3)

and the average absolute percent deviation (AAPD) is defined as

$$AAPD = \frac{1}{N} \sum |PD_i|,$$

(4)

Where V_{exp} is experimental molar volume (ft³/lbmole), V_{cal} is the calculated molar volume, and N is the number of the data points.

Table 2 shows a calculation sample of mixture number 9. The rest of the results are given in Tables 6, 7, 8, and 9. Table 3 shows the summary of statistical parameters for all mixtures. It is also shown in Table 3 the reservoir temperature, bubble point pressure, and C1 and C7+ content of each mixture. It should be noted that none of the mixtures at hand contains H_2S and they contain a little amount of CO_2 and N_2. The result of comparison reveals that none of EOSs has a grand average AAPD of less than 16. EOSs such as LLS, HK, MNM, PT and SW yield consistently high AAPD of all mixtures irrespective to their bubble point pressures and C1 and C7+ contents. The rest of EOSs (SRK, PR, and ALS) perform better for mixtures with a higher C1 content than that of a lower C1 however, with few exceptions (e.g., mixture no. 2). In the overall evaluation ALS has the least grand average AAPD (=16) among all tested EOSs.

Table 2: Calculation sample of mixture number 9

P (Psia)	Vexp (lb/lbmole)	SRK	PR	ALS	LLS	HK	MNM	PT	SW
1514.7	2.348	5.05	34.15	8.25	30.72	17.53	21.05	21.54	21.37
1491.7	2.354	5.29	34.14	8.27	30.82	17.68	21.19	21.70	21.51
1480.7	2.357	5.40	34.14	8.12	30.87	17.75	21.25	21.76	21.59
1470.7	2.361	5.13	33.40	8.07	30.85	17.76	21.23	21.76	21.59
1449.7	2.367	5.37	33.41	8.09	30.95	17.90	21.34	21.89	21.72
1439.7	2.370	5.49	33.42	8.09	31.00	17.97	21.38	21.96	21.79
1430.7	2.373	5.61	33.43	8.08	31.02	18.02	21.41	22.00	21.83
1411.7	2.379	5.84	33.45	8.09	31.10	18.15	21.53	22.11	21.95
1401.7	2.382	5.95	33.45	8.10	31.16	18.23	21.60	22.20	22.03
1383.7	2.388	6.17	33.47	8.11	31.24	18.35	21.68	22.31	22.14
1374.7	2.392	6.32	33.50	8.07	31.23	18.36	21.68	22.31	22.15
1365.7	2.395	6.42	33.51	8.07	31.28	18.43	21.73	22.38	22.23
1341.7	2.404	6.74	33.54	8.06	31.37	18.58	21.83	22.52	22.36
1332.7	2.407	6.84	33.54	8.08	31.43	18.67	21.93	22.59	22.45
1309.7	2.417	7.18	33.60	8.03	31.48	18.77	21.99	22.68	22.53
1301.7	2.420	7.27	33.61	8.04	31.53	18.84	22.04	22.74	22.60
1286.7	2.426	7.46	33.63	8.05	31.61	18.95	22.12	22.85	22.71
1272.7	2.432	7.62	33.65	8.04	31.67	19.05	22.22	22.93	22.79
1251.7	2.442	7.92	33.72	8.00	31.72	19.17	22.28	23.02	22.90
1230.7	2.451	8.22	33.75	8.03	31.85	19.35	22.43	23.20	23.06
1217.7	2.458	8.43	33.80	7.99	31.87	19.41	22.46	23.24	23.12

1199.7	2.467	8.68	33.85	7.98	31.95	19.54	22.57	23.35	23.24
1169.7	2.483	9.13	33.93	7.93	32.07	19.74	22.71	23.53	23.43
1163.7	2.486	9.21	33.94	7.94	32.11	19.80	22.76	23.58	23.52
1141.7	2.499	9.51	34.01	7.90	32.18	19.94	22.86	23.69	23.63
1114.7	2.515	10.00	34.15	7.89	32.31	20.14	23.03	23.86	23.83
1066.7	2.546	10.84	34.41	7.85	32.54	20.52	23.31	24.18	24.18
948.7	2.641	3.98	23.41	7.62	32.84	21.42	24.00	24.87	24.96
709.7	2.960	6.88	23.13	6.61	33.58	23.18	25.30	26.02	26.43
574.7	3.279	7.79	21.75	5.40	33.86	24.26	26.09	26.54	27.24
487.7	3.599	8.28	20.53	4.62	33.68	24.68	26.33	26.49	27.46
424.7	3.919	8.41	19.36	3.46	33.50	25.04	26.54	26.48	27.63
376.7	4.240	8.39	18.27	2.82	33.33	25.34	26.78	26.41	27.78
338.7	4.557	7.77	16.76	2.40	33.34	25.77	27.11	26.51	28.07
AAPD		7.19	31.00	7.36	31.88	19.89	22.82	23.39	23.47
APDmin		3.98	16.76	2.40	30.72	17.53	21.05	21.54	21.37
APDmax		7.19	31.00	7.36	31.88	19.89	22.82	23.39	23.47

Table 3: Summary of statistical parameters

Mixture no.	SRK	PR	ALS	LLS	HK	MNM	PT	SW	Data points	C1 Mol%	C7' Mol%	T F°	Pb Psig
1	11.09	9.84	9.45	30.77	23.77	26.08	27.89	26.72	10	21.05	53.73	172	1804.0
2	35.53	12.83	28.84	40.18	34.56	33.63	37.54	37.62	9	41.84	29.71	223	3366.0
3	43.01	30.02	19.10	39.46	30.85	38.33	31.02	29.73	9	1.75	88.92	234	295.0
4	33.27	20.79	19.36	50.10	37.99	42.04	39.76	40.29	15	1.45	88.03	236	344.0
5	41.73	29.03	19.67	41.29	31.32	38.20	32.04	30.83	9	1.31	93.74	231	172.0
6	25.42	13.94	11.50	42.04	29.97	33.88	31.33	31.76	14	0.93	90.25	235	266.0
7	22.27	7.76	11.56	46.09	33.69	40.33	33.72	33.31	8	1.33	92.86	247	196.0
8	22.55	3.15	27.56	47.70	37.31	38.95	40.66	40.85	11	25.37	63.87	179.3	2364.0
9	7.19	31.00	7.36	31.88	19.89	22.82	23.39	23.47	34	12.89	75.34	165	1485.3
10	5.04	25.72	8.55	26.73	15.47	18.03	18.28	18.61	23	14.67	80.72	165	1571.0
11	34.38	0.76	13.94	27.81	18.70	23.22	22.22	21.23	6	29.04	40.97	244	1970.3
Grand avr. AAPD	25.59	16.80	16.08	38.55	28.50	32.32	30.71	30.40	148*				
APDmin	5.04	0.76	7.36	26.73	15.47	18.03	18.28	18.61					
APDmax	43.01	31.00	28.84	50.10	37.99	42.04	40.66	40.85					

Total* number of data points.

The reported inaccuracy is not peculiar to Sudanese reservoir fluids but it is rather a known problem associated with EOSs. Many investigators such as Coats and Smart [19] and Ahmed [20], to mention a few, have reported the inaccuracy of EOSs to reservoir fluids. The inaccuracy can be attributed to the following EOSs limitations and plus fraction properties.

- High level of uncertainty in the prediction of the critical properties and the accentric factor of plus fraction which are not measured in the laboratory. Bearing in mind that plus fraction is the main constituent of mixtures. It constitutes more than 75% in most of the mixtures at hand (mixtures no. 3 to no. 7 and mixture no. 9). The inaccuracy in prediction of critical properties and acentric factor is due to the fact that the plus fraction lumps millions of compounds that only few of them (C7 to C36) are known by measurement. Hence it is expected that as the content of the plus fraction in the mixture increases the inaccuracy of EOS increases. This may justify the relatively good performance of SRK, PR and ALS for mixtures of a lower percentage of plus fraction.

- The parameters of the attraction term α, $\alpha(T)$ and covolume b of EOS are determined based on van der Waal critical point assumption while the reservoirs temperatures in this work (cf. Table 1) are higher than the critical temperatures of N_2, CO_2, C1, and C2. For mixtures numbers 1, 2, 8, and 11, these components together constitute more than 20% of the said mixture. This means that the applicability limit of EOS is violated and hence the inaccuracy of EOSs is not a surprise.

- Lack of information on BIPs of N_2–CO_2, N_2-hydrocarbon, and CO_2-hydrocarbon, However, the presence of the nonhydrocarbon components (N_2 and CO_2) is small in the investigated mixtures.

- All EOSs used in this work contain three [$\alpha,b,\alpha(T)$] and four [$\alpha,b,\alpha(T),b_1$ and b_2 or c] parameters, hence besides the van der Waal critical point conditions, the parameters were determined by regressing experimental data for pure components. The commonly used experimental data for such a purpose include vapor pressure, normal boiling point and density at standard conditions (T = 15°C and P= 1 atm). These data are generally for lower molecular weight components. Hence an EOS that developed on these data unlikely will suffice for reservoir fluid which contains a higher molecular weight plus fraction.

Because the inaccuracy of EOSs rests on the four reasons outlined above, a number of methods have been proposed over years to enhance the capability of EOS yet maintaining its original characteristics. These accuracy enhancement methods include the following.

- Development of accurate models for the prediction of critical properties and acentric factor of plus fraction.
- Application of a volume-translation technique such as Peneloux shift factor to the EOS.
- Tuning EOS to experimental data. However, Pedersen [21] warned that "using equation of state parameters "tuned" to one specific property yields unreliable predictions of other thermodynamic properties."
- Modification of the expression used in the denominator of the attractive term Yu et al. [22]. This method as mentioned earlier is employed by Adepoju [1]. He has made a significant improvement in the accuracy of PR and SRK by replacing their parameters by those of LLS.

In this work ALS EOS which produced the least grand average AAPD (=16) of all mixtures is considered as the candidate to predict PVT data of the Sudanese reservoir fluids. Enhancement procedure, using the technique number 4 listed above, is considered.

Prior to the employment of the modification, ALS is described as

$$P = \frac{RT}{v - b_1} - \frac{a(T)}{(v - b_2)(v + b_3)},$$

$$b_i = \frac{\Omega_{b_i} R T_c}{P_c}, \quad i = 1, 2, 3,$$

$$a(T) = \frac{\Omega_a (R T_c)^2 \alpha(T)}{P_c},$$

(5)

$$\Omega_a = 0.44869 + 0.04024\omega + 0.01111\omega^2 - 0.00576\omega^3,$$

(6a)

$$\Omega_{b1} = 0.08974 - 0.03452\omega + 0.00330\omega^2, \tag{6b}$$

$$\Omega_{b2} = 0.5\left[2(1 + \Omega_{b1}) - 3\Omega_a^{1/3} + \left(4\Omega_a - 3\Omega_a^{2/3}\right)^{1/2}\right], \tag{6c}$$

$$\Omega_{b3} = 0.5\left[-2(1 + \Omega_{b1}) + 3\Omega_a^{1/3} + \left(4\Omega_a - 3\Omega_a^{2/3}\right)^{1/2}\right]. \tag{6d}$$

The modification includes the replacement of the first term of the coefficient Ωb_1 (cf. (6b)) by that of PR EOS ($\Omega b = 0.07780$). Hence the modified form of Ωb_1 is

$$\Omega_{b1} = \underbrace{0.07780}_{PR} - 0.03452\omega + 0.00330\omega^2. \tag{7}$$

Remember that the modification of Ωb_1 will automatically modify the coefficients Ωb_2 and. Ωb_3.

Table 4 shows calculation samples of mixtures numbers 1 and 6. Table 5 shows the summary of AAPD for all mixtures after the modification of the ALS parameters. It can be seen that the accuracy of ALS enhanced by a factor of 30 to 90%. The grand average AAPD is reduced significantly from 16 to 7.

Table 4: Calculation samples of mixtures numbers 1 and 6 with ALS in its original and modified forms

Mixture no. 1				Mixture no. 6			
P	Vexp	AAPD		P	Vexp	AAPD	
Psia	lb/lb mole	Original	Modified	Psia	lb/lb mole	Original	Modified
1818.7	2.548	−13.08	−0.18	280.7	4.532	−16.76	−3.17
1765.7	2.573	−13.02	−0.24	279.7	4.538	−16.76	−3.19
1750.7	2.579	−13.06	−0.31	278.7	4.545	−16.73	−3.19
1473.7	2.763	−12.07	−0.17	276.7	4.556	−16.77	−3.25
1177.7	3.086	−10.41	0.25	271.7	4.592	−16.66	−3.25
993.7	3.411	−8.90	0.74	267.7	4.621	−16.59	−3.26
865.7	3.736	−7.56	1.24	233.7	4.918	−15.13	−2.61
769.7	4.062	−6.41	1.69	209.7	5.215	−13.90	−2.09
694.7	4.388	−5.42	2.08	175.7	5.810	−11.42	−0.82
634.7	4.711	−4.57	2.41	153.7	6.404	−8.60	1.01

				136.7	6.999	−6.16	2.64
				123.7	7.593	−3.82	4.29
				113.7	8.188	−1.50	6.02
				104.7	8.784	0.26	7.27
	AAPD	9.45	0.93	207.63	5.81	11.50	3.29
	APDmin	4.57	0.17	104.70	4.53	0.26	0.82
	APDmax	13.08	2.41	280.70	8.78	16.77	7.27

Table 5: AAPD of ALS after modification

Mixtures											
Statistical parameters	1	2	3	4	5	6	7	8	9	10	11
AAPD	0.93	19.68	7.51	7.53	6.14	3.29	6.50	15.54	5.16	5.53	1.55
APDmin	0.17	17.50	1.18	1.22	4.87	0.82	0.46	14.34	4.30	3.29	0.04
APDmax	2.41	20.85	10.81	9.29	8.23	7.27	13.44	16.54	5.60	13.24	3.00
Enhancement %	90.17	31.75	60.66	61.08	68.81	71.39	43.80	43.61	29.95	35.32	88.91

Table 6: PD of mixtures numbers 2 and 3

P(psia)	Vexp (ft3/lbmole))	SRK	PR	ALS	LLS	HK	MNM	PT	SW
Mixture no. 2									
3380.7	1.787	-50.38	-8.17	-30.91	-43.78	-32.05	-34.19	-36.03	-35.68
3181.7	1.807	-49.59	-10.00	-32.37	-45.13	-33.58	-35.52	-37.68	-37.31
2234.7	2.129	-40.75	-12.61	-32.11	-43.41	-34.80	-34.63	-38.65	-38.24
1801.7	2.446	-36.05	-13.29	-30.54	-41.34	-35.00	-33.75	-38.29	-38.26
1531.7	2.765	-32.73	-13.73	-28.98	-39.54	-34.92	-33.07	-37.86	-38.01
1339.7	3.085	-30.21	-13.74	-27.63	-38.23	-34.88	-32.70	-37.51	-37.81
1194.7	3.405	-28.27	-13.79	-26.52	-37.23	-34.82	-32.48	-37.19	-37.59
1079.7	3.725	-26.86	-13.93	-25.66	-36.53	-34.81	-32.40	-36.95	-37.43
764.7	5.078	-24.96	-16.16	-24.81	-36.41	-36.16	-33.88	-37.65	-38.26
Mixture no. 3									
309.7	6.119	-51.65	-35.61	-22.17	-46.35	-36.18	-45.00	-36.33	-34.81
299.7	6.164	-51.82	-36.00	-22.70	-46.72	-36.23	-45.38	-36.77	-35.27
290.7	6.209	-52.07	-36.32	-23.13	-47.03	-36.68	-45.70	-37.15	-35.28
272.7	6.317	-52.14	-36.84	-23.95	-47.48	-37.30	-46.16	-37.36	-35.94
233.7	6.666	-51.10	-36.97	-24.94	-47.32	-37.68	-45.65	-37.74	-36.39
197.7	7.228	-47.75	-34.92	-23.99	-44.24	-35.92	-43.28	-35.99	-34.74
159.7	8.344	-39.27	-28.45	-19.20	-36.86	-29.63	-36.01	-29.70	-28.62
130.7	10.009	-27.17	-18.27	-10.70	-25.63	-19.55	-24.89	-19.63	-18.72
109.7	12.144	-14.08	-6.82	-1.12	-13.50	-8.48	-12.88	-8.54	-7.80

Table 7: PD of mixtures numbers 4 and 5

P(psia)	Vexp (ft3/lbmole)	SRK	PR	ALS	LLS	HK	MNM	PT	SW
Mixture no. 4									
359.7	4.082	-40.11	-25.08	-23.04	-54.75	-41.00	-45.76	-43.22	-43.62
357.7	4.092	-40.00	-25.03	-23.00	-54.70	-40.97	-45.72	-43.19	-43.59
357.7	4.092	-40.00	-25.03	-23.00	-54.70	-40.97	-45.72	-43.19	-43.59
355.7	4.102	-39.90	-24.98	-22.96	-54.66	-40.96	-45.69	-43.17	-43.56
352.7	4.113	-39.89	-25.03	-23.03	-54.75	-41.07	-45.80	-43.27	-43.68
350.7	4.124	-39.76	-24.95	-22.97	-54.67	-41.03	-45.74	-43.22	-43.63
344.7	4.152	-39.66	-24.88	-22.96	-54.68	-41.09	-45.77	-43.26	-43.68
334.7	4.206	-39.16	-24.65	-22.79	-54.48	-41.03	-45.65	-43.16	-43.60
291.7	4.479	-36.95	-23.63	-21.70	-54.06	-40.82	-45.18	-42.77	-43.28
235.7	5.026	-32.82	-21.14	-19.77	-51.48	-38.94	-42.77	-40.56	-41.16
200.7	5.574	-28.75	-18.35	-17.40	-47.58	-36.80	-40.27	-38.19	-38.85
175.7	6.122	-25.13	-15.82	-15.22	-44.58	-34.64	-37.81	-35.85	-36.54
156.7	6.669	-22.10	-13.60	-13.20	-41.91	-32.67	-35.60	-33.74	-34.44
142.7	7.218	-18.94	-11.11	-10.85	-38.83	-30.21	-32.92	-31.16	-31.87
131.7	7.766	-15.87	-8.65	-8.54	-35.72	-27.65	-30.18	-28.51	-29.21
Mixture no. 5									
186.7	6.026	-46.53	-31.69	-20.37	-45.12	-33.43	-41.69	-34.61	-33.20
184.7	6.044	-46.49	-31.72	-20.45	-45.13	-33.48	-41.72	-34.65	-33.25
180.7	6.081	-46.43	-31.79	-20.61	-45.16	-33.58	-41.77	-34.75	-33.35
175.7	6.131	-46.32	-31.85	-20.80	-45.17	-33.68	-41.80	-34.84	-33.46
166.7	6.232	-46.13	-31.87	-21.01	-45.12	-33.94	-41.14	-34.36	-33.06

144.7	6.551	-44.98	-31.62	-21.47	-44.52	-33.87	-40.72	-34.27	-33.03
106.7	7.527	-40.44	-29.13	-20.64	-40.14	-31.64	-37.63	-32.01	-30.93
77.7	9.086	-32.75	-23.45	-17.51	-34.16	-27.01	-32.00	-27.34	-26.43
57.7	11.223	-25.52	-18.14	-14.21	-27.06	-21.25	-25.30	-21.52	-20.79

Table 8: PD of mixtures numbers 7 and 8

P(psia)	Vexp (ft3/lbmole)	SRK	PR	ALS	LLS	HK	MNM	PT	SW
Mixture no. 7									
210.7	5.73	-23.11	-5.38	-7.24	-45.58	-31.02	-39.24	-31.09	-30.53
196.7	5.839	-22.64	-5.35	-7.90	-46.00	-31.75	-39.23	-31.81	-31.29
190.7	5.893	-22.25	-5.11	-8.18	-46.13	-32.03	-39.46	-32.08	-31.58
179.7	6.001	-22.05	-5.43	-8.76	-46.39	-32.53	-39.85	-32.58	-32.10
153.7	6.325	-21.36	-5.93	-9.71	-46.28	-33.78	-40.78	-33.81	-33.39
115.7	7.098	-21.24	-8.03	-12.93	-46.94	-35.47	-41.76	-35.47	-35.14
86.7	8.185	-21.36	-10.37	-16.35	-46.62	-36.55	-42.01	-36.53	-36.27
52.7	11.076	-24.15	-16.47	-21.43	-44.79	-36.42	-40.32	-36.43	-36.18
Mixture no. 8									
2310.7	1.995	-31.27	3.74	-29.60	-49.18	-34.73	-38.06	-39.07	-38.90
2251.7	2.007	-30.73	3.94	-29.57	-49.19	-34.90	-38.15	-39.23	-39.05
2036.7	2.058	-28.73	4.38	-29.50	-49.24	-35.58	-38.49	-39.88	-39.66
1867.7	2.11	-27.01	4.51	-29.34	-49.16	-36.06	-38.68	-40.31	-40.10
1563.7	2.24	-23.89	4.05	-28.81	-48.85	-36.95	-39.00	-41.01	-40.88
1202.7	2.502	-20.34	2.43	-27.94	-48.09	-37.93	-39.24	-41.53	-41.57

988.7	2.764	-18.56	0.67	-27.09	-47.21	-38.43	-39.33	-41.58	-41.83
842.7	3.028	-17.58	-0.91	-26.35	-46.66	-38.76	-39.40	-41.49	-41.93
737.7	3.292	-16.89	-2.17	-25.59	-46.06	-38.84	-39.30	-41.19	-41.80
655.7	3.556	-16.67	-3.44	-25.10	-45.77	-39.09	-39.42	-41.10	-41.87
593.7	3.812	-16.39	-4.38	-24.30	-45.33	-39.09	-39.36	-40.84	-41.73

Table 9: PD of mixtures numbers 10 and 11

P(psia)	Vexp (ft3/lbmole)	SRK	PR	ALS	LLS	HK	MNM	PT	SW
Mixture no. 10									
1585.7	2.406	-1.38	27.91	-10.65	-33.03	-18.90	-22.33	-22.57	-22.78
1565.7	2.41	-1.24	27.85	-10.71	-33.15	-19.05	-22.47	-22.72	-22.93
1552.7	2.415	-1.05	27.88	-10.64	-33.11	-19.04	-22.44	-22.71	-22.93
1540.7	2.42	-0.86	27.91	-10.43	-33.05	-19.01	-22.40	-22.67	-22.89
1528.7	2.424	-0.73	27.91	-10.41	-33.04	-19.04	-22.41	-22.69	-22.91
1499.7	2.435	-0.35	27.94	-10.31	-32.98	-19.05	-22.39	-22.69	-22.91
1471.7	2.446	0.02	27.97	-10.21	-32.92	-19.07	-22.38	-22.70	-22.92
1445.7	2.457	0.38	28.00	-10.10	-32.85	-19.07	-22.35	-22.69	-22.91
1369.7	2.492	1.42	28.10	-9.78	-32.62	-19.03	-22.24	-22.62	-22.84
1326.7	2.514	2.05	28.17	-9.58	-32.45	-19.00	-22.15	-22.58	-22.79
1247.7	2.558	3.23	28.36	-9.25	-32.17	-18.98	-22.04	-22.50	-22.64
1275.7	2.542	2.79	28.27	-9.35	-32.26	-18.98	-22.07	-22.52	-22.73
1270.7	2.545	2.94	28.35	-9.33	-32.23	-18.97	-22.05	-22.51	-22.72
1246.7	2.559	3.24	28.38	-9.23	-32.14	-18.96	-22.02	-22.50	-22.70

1211.7	2.582	3.80	28.49	-9.02	-31.95	-18.88	-21.89	-22.38	-22.60
1176.7	2.605	4.36	28.62	-8.86	-31.81	-18.87	-21.83	-22.33	-22.55
777.7	3.063	7.56	24.79	-5.72	-27.78	-17.10	-19.43	-20.03	-20.31
439.7	4.218	11.15	21.87	-0.67	-20.27	-12.44	-13.85	-14.17	-14.58
311.7	5.379	12.00	19.81	2.66	-15.37	-8.99	-9.87	-9.99	-10.53
243.7	6.536	12.81	18.99	4.86	-11.33	-6.34	-7.14	-7.03	-7.69
201.7	7.695	13.70	18.82	6.87	-8.23	-3.95	-4.66	-4.38	-5.13
171.7	8.854	14.04	18.42	8.13	-6.17	-2.44	-3.07	-2.65	-3.49
150.7	10.012	14.78	18.63	9.87	-3.97	-0.68	-1.25	-0.71	-1.62
Mixture no. 11									
1985	3.137	-39.42	0.28	-13.20	-27.27	-18.19	-22.30	-20.55	-19.42
1860	3.210	-37.26	-0.04	-13.54	-27.45	-18.80	-22.58	-21.13	-20.04
1728	3.304	-34.90	-0.69	-13.83	-27.63	-18.80	-22.92	-21.81	-20.78
1617	3.398	-33.02	-0.93	-14.09	-27.86	-18.80	-23.33	-22.52	-21.57
1523	3.489	-31.54	-1.22	-14.39	-28.19	-18.80	-23.86	-23.30	-22.42
1441	3.583	-30.15	-1.43	-14.59	-28.44	-18.80	-24.32	-23.98	-23.17

CONCLUSIONS

The work provided important information on PVT data on Sudanese reservoir fluids. It includes composition analysis of fraction plus up to $C7^+$ and about 148 data points of CME test (pressure-volume data) at pressures below the bubble point. The paper presents also a modified form of ALS that describes the Sudanese reservoir fluids with a good level of accuracy.

ACKNOWLEDGMENTS

The authors acknowledge the support of the Ministry of Energy and Mining, Sudan.

REFERENCES

1.   O. O. Adepoju, Coefficient of isothermal oil compressibility for reservoir fluids by cubic equation of state, M.Sc. thesis, University of Texas, Austin, Tex, USA, 2006.

2.   D.-Y. Peng and D. B. Robinson, "A new two-constant equation of state," Industrial and Engineering Chemistry Fundamentals, vol. 15, no. 1, pp. 59–64, 1976.

3.   G. Soave, "Equilibrium constants from a modified Redlich-Kwong equation of state," Chemical Engineering Science, vol. 27, no. 6, pp. 1197–1203, 1972.

4.   A. S. Lawal, E. T. Van der Laan, and R. K. M. Thambynayagam, "Four-parameter modification of the Lawal-Lake-Silberberg equation of state for calculating gas-condensate phase equilibria," inProceedings of the Annual Technical Conference and Exhibition, Las Vegas, Nev, USA, September 1985, paper SPE 14269.

5.   K. Akbarzadeh, Sh. Ayatollahi, Kh. Nasrifar, H. W. Yarranton, and M. Moshfeghian, "Prediction of the densities of Western Canadian heavy oils and their SARA fractions from the cubic equations of state," Iranian Journal of Science and Technology, Transaction B, vol. 28, no. B6, pp. 695–699, 2004.

6.   B. H. Jensen, Densities, viscosities and phase equilibria in enhanced oil recovery, Ph.D. thesis, Department of Chemical Engineering, the Technical University of Denmark, Lyngby, Denmark, 1987.

7.   Y. Adachi, B. C.-Y. Lu, and H. Sugie, "A four-parameter equation of state," Fluid Phase Equilibria, vol. 11, no. 1, pp. 29–48, 1983.

8.   K. Aasberg-Petersen, Bulk phase properties and phase equilibria for miscible and immiscible oil displacement processes, Ph.D. thesis progress report, Department of Chemical Engineering, the Technical University of Denmark, Lyngby, Denmark, 1989.

9.   A. Peneloux, E. Rauzy, and R. Freze, "A consistent correction for Redlich-Kwong-Soave volumes,"Fluid Phase Equilibria, vol. 8, no. 1, pp. 7–23, 1982.

10.   K. Aasberg-Petersen and E. Stenby, "Prediction of thermodynamic properties of oil and gas condensate mixtures," Industrial and Engineering Chemistry Research, vol. 30, no. 1, pp. 248–254, 1991.

11.   R. A. Almehaideb, I. Ashour, and K. A. El-Fattah, "Improved K-value correlation for UAE crude oil components at high pressures using PVT laboratory data," 2003.

12.   M.-L. Yu and Y.-P. Chen, "VLE calculations by applying a modified perturbed hard sphere EOS,"Fluid Phase Equilibria, vol. 129, no. 1-2, pp. 21–35, 1997.

13.   N. Kumer, Compressibility factor for natural and sour reservoir gases by correlations and cubic equations of state, M.Sc. thesis, Texas Tech University, Lubbock, Tex, USA, 2004.

14.   A. A. Rabah and S. A. Mohamed, "Prediction of molar volumes of undersaturated Sudanese reservoir fluids," submitted to Journal of Petroleum Science and Engineering.

15.   A. Danesh, PVT and Phase Behavior of Petroleum Reservoir Fluids, Elsevier Science, Amsterdam, The Netherlands, 2nd edition, 1998.

16.   A. O. Tododo, Thermodynamically equivalent pseudo components for compositional reservoir simulation models, Ph.D. dissertation, Texas Tech University, Lubbock, Tex, USA, 2005.

17.   A. Rabah and S. Kabelac, "Flow boiling of R134a and R134a/ propane mixtures at low saturation temperatures inside a plain

horizontal tube," Journal of Heat Transfer, vol. 130, no. 6, Article ID 061501, 9 pages, 2008.

18. C. H. Whitson and M. R. Brule', Phase Behavior, vol. 20 of Monograph: SPE Henry L. Doherty Series, SPE, Richardson, Tex, USA, 2000.

19. K. H. Coats and G. T. Smart, "Application of a regression-based EOS PVT program to laboratory data," SPE Reservoir Engineering, vol. 1, no. 3, pp. 277–299, 1986.

20. T. Ahmed, "On equation of state," in Proceedings of the SPE Latin American and Caribbean Petroleum Engineering Conference, vol. 1, pp. 539–559, Buenos Aires, Argentina, 2007.

21. K. S. Pedersen, "On the dangers of "Tuning" equation of state parameter," Society of Petroleum Engineers of AIME, p. 14487, 1985.

22. J. M. Yu, S. H. Huang, and M. Radosz, "Phase behavior of reservoir fluids: supercritical carbon dioxide and cold lake bitumen," Fluid Phase Equilibria, vol. 53, pp. 429–438, 1989.

Chapter 2

Field Scale Simulation Study of Miscible Water Alternating CO$_2$ Injection Process in Fractured Reservoirs

Mohammad Afkhami Karaei, Ali Ahmadi,
Hooman Fallah, Shahrokh Bahrami Kashkooli,
and Jahangir Talebi Bahmanbeglo

Department of Petroleum Engineering, Islamic Azad University, Firoozabad Branch, Firoozabad, Iran

ABSTRACT

Vast amounts of world oil reservoirs are in natural fractured reservoirs. There are different methods for increasing recovery from fractured reservoirs. Miscible injection of water alternating CO$_2$ is a good choice among EOR methods. In this method, water and CO$_2$ slugs are injected alternatively in reservoir as miscible agent into reservoir. This paper studies water injection scenario and miscible injection of water and CO$_2$ in a two dimensional, inhomogeneous fractured reservoir. The results show that miscible water alternating CO_2 gas injection leads to 3.95% increase in final oil recovery and total water production decrease of 3.89% comparing to water injection scenario.

INTRODUCTION

Miscible gas injection is one of the most important mechanisms of enhanced oil recovery from fractured reservoirs. Miscible gas injection may liberate and lead to production of a lot of amounts of oil that is entrapped into matrixes. Recovery from fracture-matrix system in laboratory scales started from 1970. Thompson and Mungan investigated results of laboratory experiments of gravity drainage in a fractured porous media under first contact miscible conditions. Basically, they compared replacement velocity with critical velocity and investigated its effect on oil recovery factor [1] - [5]. Water alternating gas injection is an effective method for controlling high mobility of gas in horizontal flooding which is used in so many reservoirs around the world and reported as successful operation. On the other hand, immiscible water alternating gas injection in oil reservoirs is not common [2]. Application of immiscible water alternating gas injection in USA is usually limited to onshore reservoirs. There are so many gas types with different characterizations used in miscible injection. Although many miscible gases include CO_2 and hydrocarbon gases, abundance of CO_2 and technical availability is an important parameter for improving miscible gas injection in USA [3]. Oil recovery factor using water alternating CO_2 gas injection is more than water injection, CO_2 injection, and hot CO_2 injection. Because of gas mobility, it penetrates into places which are unavailable during common injections [4].

Water flooding is an EOR method. One of the problems of water flooding is that the ions in water may be incompatible with reservoir fluid. It can lead to problems in ionic equilibrium which causes precipitation of heavy components of oil that can be a serious challenge in recovery from reservoir. Miscible water alternating CO_2 solves this inconvenience. Purpose of this study is to compare water injection and water alternative CO_2 gas injection and investigation of its effect on recovery factor and water cut from reservoir.

INTERPRETING RESERVOIR MODEL

Simulation of different phases in this project is done using a commercial software. Water flooding and water alternating CO$_2$ gas is tested in a two dimensional 1 × 25 × 25 grid block network.

A reservoir with all of no-flow boundaries are studied in this project. There are two phases of water and oil and no free gas. This model indicates a 200 acre (about 2950 ft × 2950 ft) reservoir. There is an injection well named INJ at the center of reservoir (at 13:13:1 grid block) and four production wells at the corners named P1, P2, P3, and P4. Production well P1 is at 1:1:1 grid block, P2 at 25:1:1, P3 at 1:25:1, and P4 is placed at 25:25:1.

Figure 1 depicts this reservoir from above. The wells are drilled with a 40 acre distance from each other and all of the wells started to production simultaneously. Depth from top surface of reservoir is 10,000 ft and reservoir net pay thickness is 50 ft.

Basic geological characteristics and different rock properties (porosity, absolute permeability, etc.) in each grid are specified at center of the grid block. Empty space volumes of blocks and inter-block transmissibility are calculated via simulator. The keywords used in this part depend on selected geometry option in initialization section. Cartesian, block-centered geometry option is used in this project. Porosity distribution is assumed to be homogeneous in reservoir and its value is 25%. But, permeability is inhomogeneous and has an average amount of 60 mD for the basic case. Original fluids in place including water and oil are saturated. Water occupies 20 percent of empty volumes and oil 80 percent. Residual oil and connate water are 15% and 20%, respectively. Initial pressure of reservoir is 4500.

As explained before, all the wells are drilled vertically. Inside diameter of wells are 0.5 ft and their depth is 10,050 ft and all the wells started production simultaneously (1st January 2010). All the wells produced for a period of 10 years with a constant 6 month controllers. There are 20 controller for each well in production period. Injection program has a controlled rate. Injected water rate is 3500 STB/day and injection rate of water alternating CO$_2$ is 3500 STB/day and 3000 Cuf/day, respectively. Production wells bottom-hole pressures are considered as constraints in water flooding. Minimum acceptable bottom-hole pressure is considered 2500 psi.

PVT PROPERTIES OF RESERVOIR FLUIDS

Reservoir fluids are water and oil. The oil contains a constant and homogeneous saturation of dissolved gas with amount of 0.2 MSTB/day. Oil bubble pressure assumed 400 psi. Oil viscosity in base pressure of 4500 psi is 2.4 cp. Oil formation volume factor is 0.972. At the condition that water density assumed 2.4 lb/cuft, oil density would be 56 lb/cuft. Water compressibility assumed 3×10^{-6} psi^{-1}. At basic pressure of 4500 psi water formation volume factor assumed 1.0034 rb/stb and water viscosity assumed 0.96 cp. Rock compressibility assumed 1.4×10^{-6} psi^{-1}.

WATER INJECTION AND MISCIBLE WATER ALTERNATING CO_2 GAS INJECTION SIMULATION

First, permeability distribution is depicted in Figure 2.

Cumulative oil production and water cut decrease during water flooding and miscible water alternating CO_2 gas flooding is shown in Figure 3 and Figure 4, respectively. After miscible water alternating CO_2 gas injection, cumulative produced oil increases 3.95 percent and cumulative produced water decreases 3/89 percent. Effect of miscible water alternating CO_2 gas injection on water saturation is shown in Figure 5 and Figure 6.

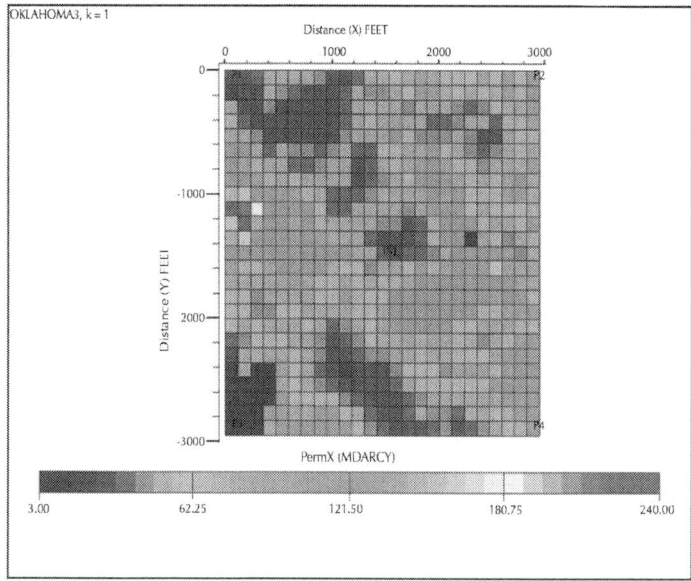

Figure 1: View of model from above that show permeability distribution in field and places of wells.

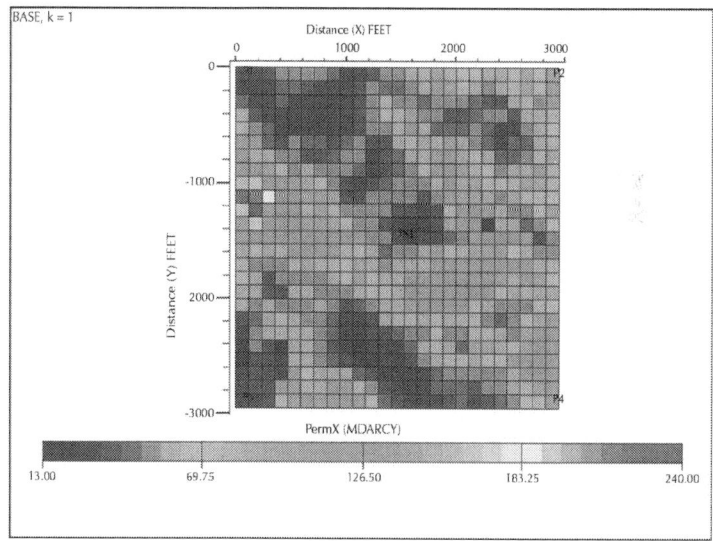

Figure 2: Assumed reservoir permeability distribution.

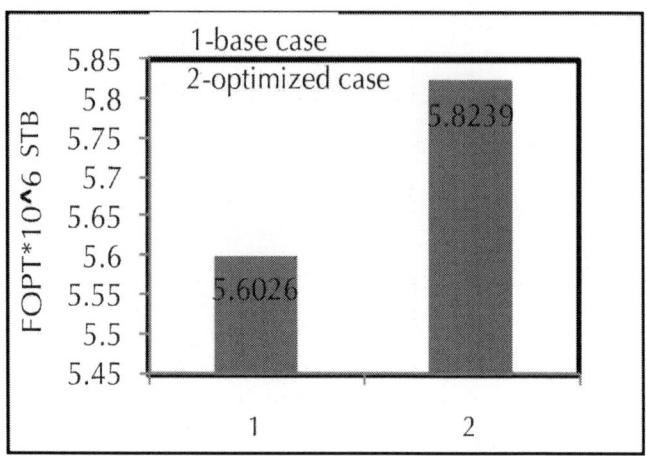

Figure 3: Cumulative produced oil in water flooding (1) and miscible water alternating CO_2 gas injection (2).

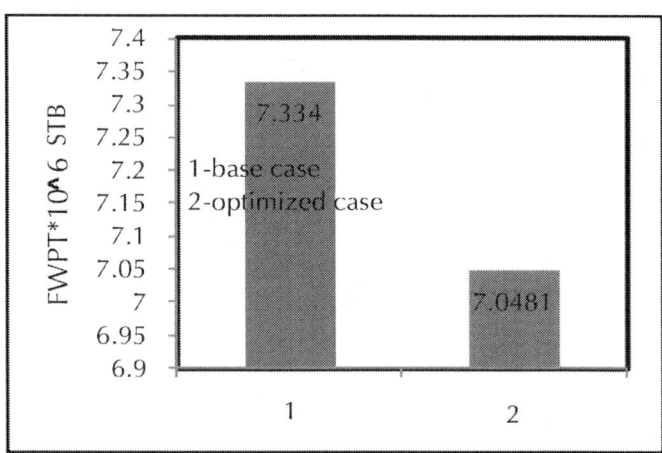

Figure 4: Cumulative produced water in water flooding (1) and miscible water alternating CO_2 gas injection (2).

There is a noticeable change in water saturation distribution in all over the reservoir before the breakthrough. This can be seen in Figure 5 and Figure 6. Figure 5 shows water saturation distribution during water injection and Figure 6 shows water saturation distribution after miscible water alternating CO_2 gas injection. Via comparing Figure 5 and Figure 6 it is obvious that water saturation is more homogeneously distributed all thorough the reservoir in Figure 5. This means that sweeping efficiency increases after miscible water alternating CO_2 gas injection.

Figure 7 shows water cut for all the wells for water injection and Figure 8 shows water cut for all the wells in miscible water alternating CO_2 gas injection.

Water cuts of four wells are shown in Figure 7 and Figure 8 before and after miscible water alternating CO_2 gas injection. Via comparing Figure 7 and Figure 8 it is obvious that water breakthrough time and water cut curves almost tend to cover each other.

Water production rate curve for each production well during water flooding and miscible water alternating CO_2 gas injection also are shown in Figures 9-12.

By comparing above figures it can be seen that water production rate of production wells P1 and P3 increases after miscible water alternating CO_2 gas injection and water breakthrough happens earlier. It is because of bottom-hole pressure decrease of production wells P1 and P2 and oil production rate increase from these wells. Also, it is obvious that water production decreases from production wells P2 and P4 after miscible water alternating CO_2 gas injection and water breakthrough time is delayed. It is because of bottom-hole pressure increase of production well P2 and P4 and decrease in produced oils of these wells. Cumulative produced oil and water vs. time curves during water injection and miscible water alternating CO_2 gas injection are shown in Figure 13.

Cumulative produced oil and water plots are shown above. Via comparing these curves it is obvious that cumulative produced oil is increased after miscible water alternating CO_2 gas injection and cumulative produced water is decreased. In this project, cumulative produced oil is increased 3.95% and cumulative produced water is decreased 3.89.

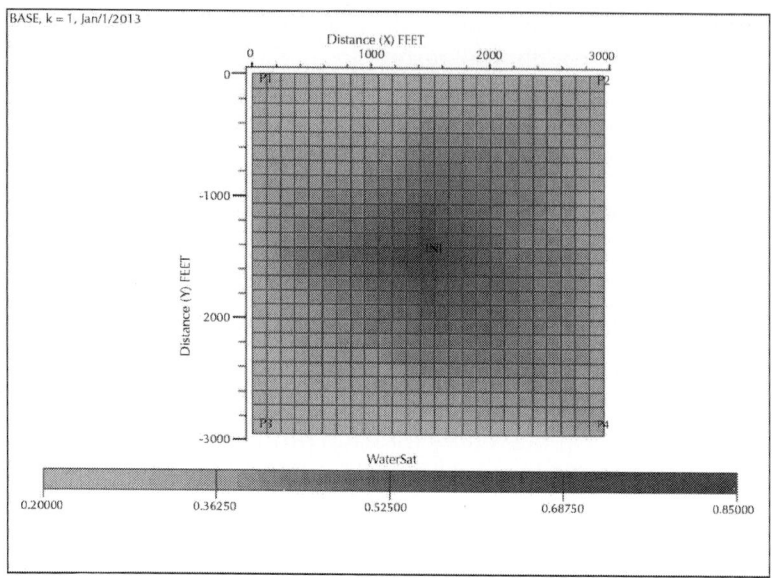

Figure 5: Water saturation distribution in water flooding.

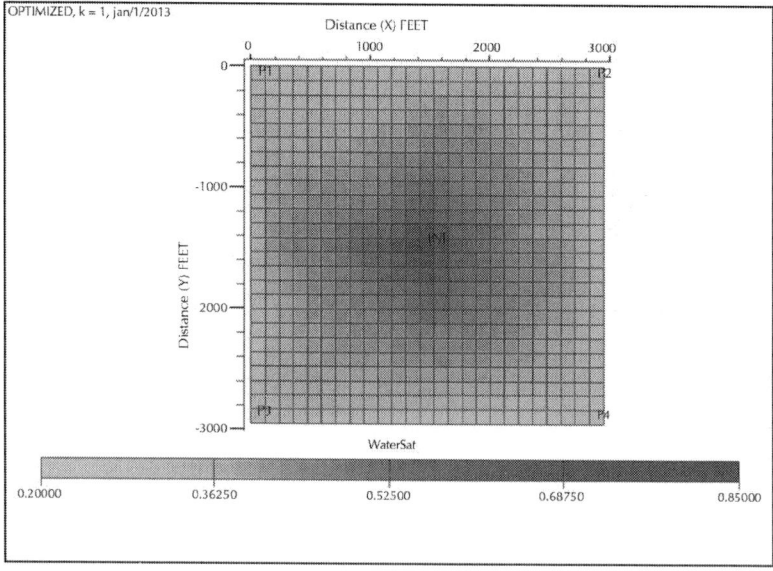

Figure 6: Water saturation distribution in miscible water alternating CO_2 gas injection.

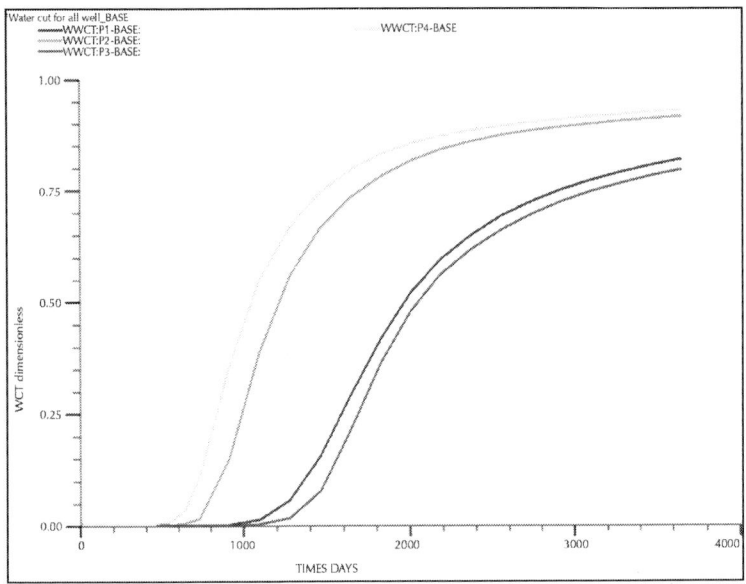

Figure 7: Water cut curves for all the wells during water injection.

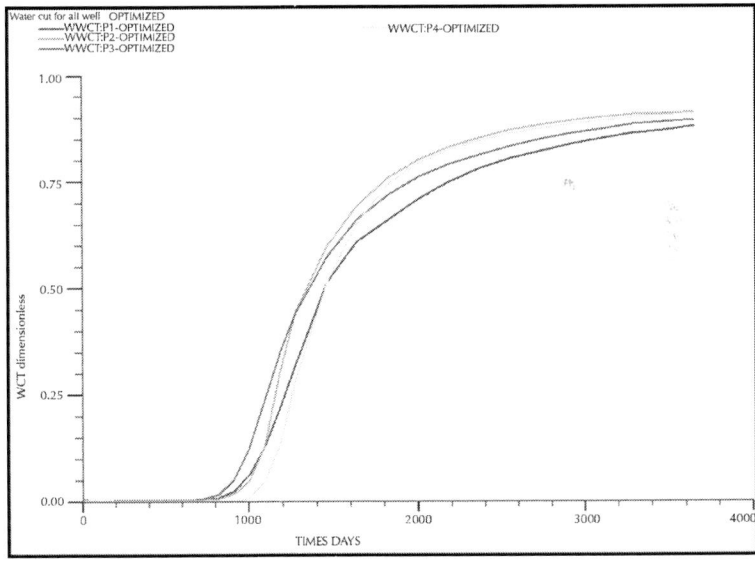

Figure 8: Water cut curves for all the wells during miscible water alternating CO$_2$ gas injection.

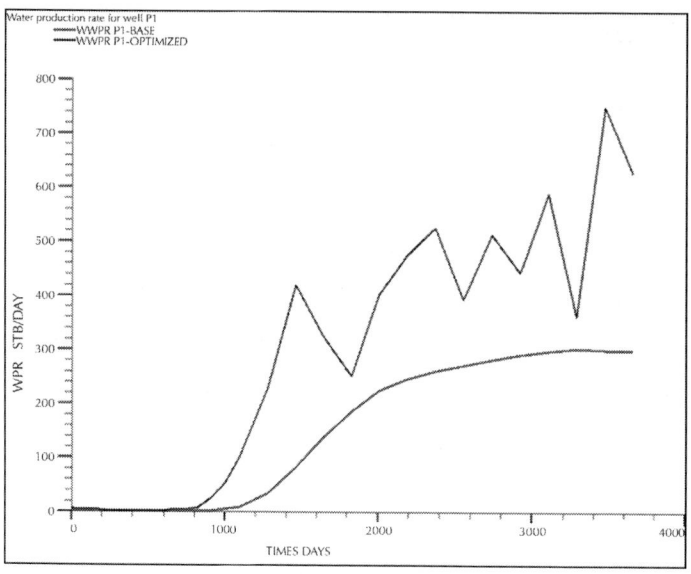

Figure 9: Water production rate from well P1 during water injection and miscible water alternating CO_2 gas injection.

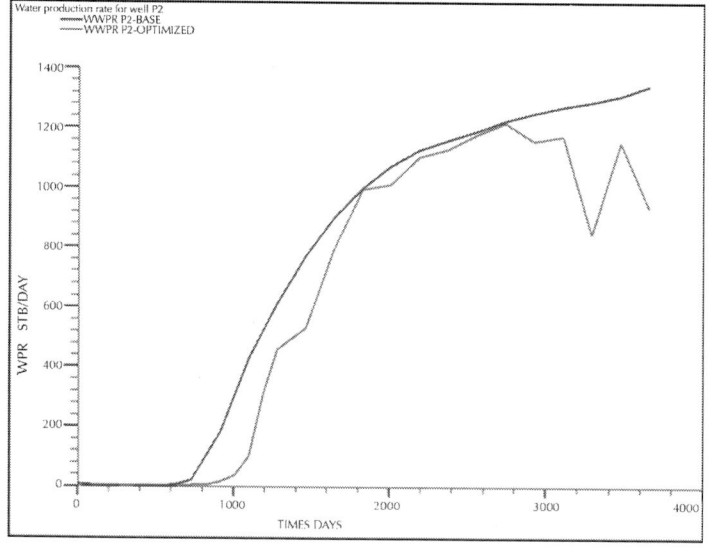

Figure 10: Water production rate from well P2 during water injection and miscible water alternating CO_2 gas injection.

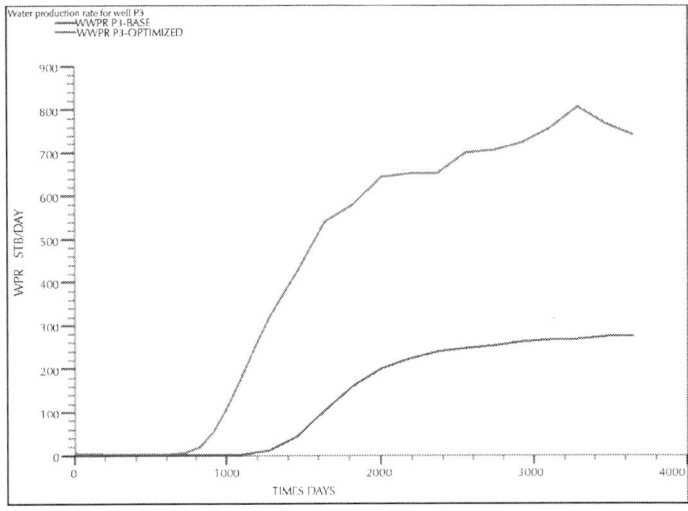

Figure 11: Water production rate from well P3 during water injection and miscible water alternating CO$_2$ gas injection.

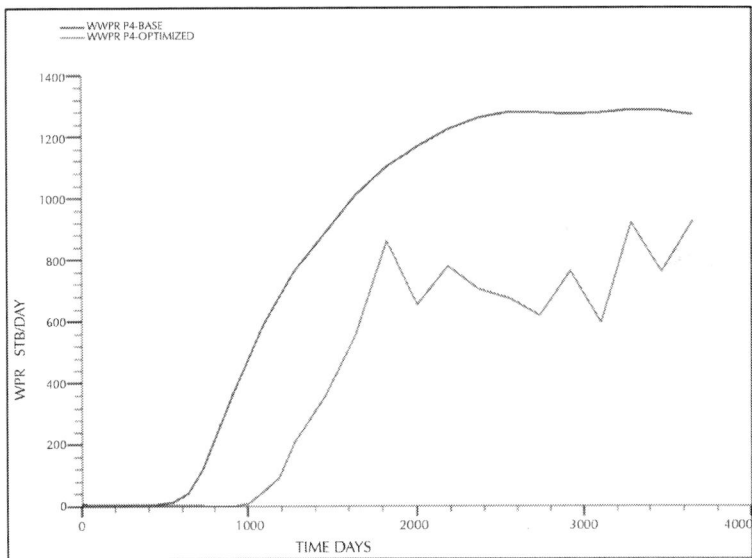

Figure 12: Water production rate from well P4 during water injection and miscible water alternating CO$_2$ gas injection.

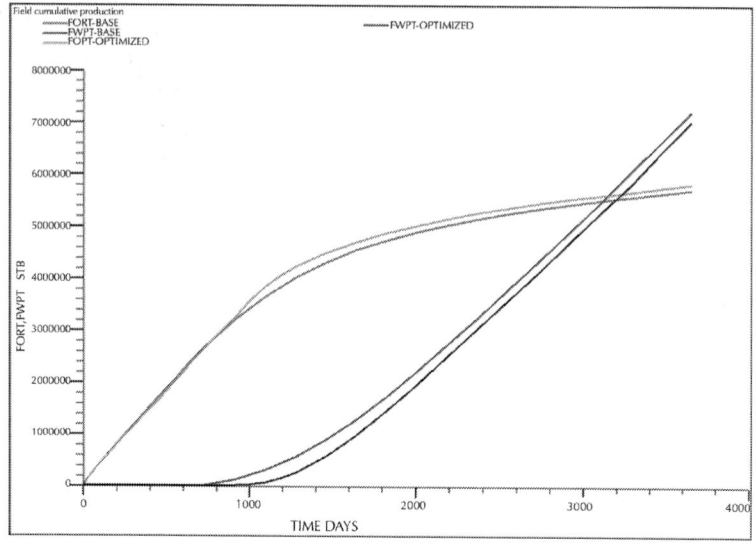

Figure 13: Cumulative produced oil and water during water injection and miscible water alternating CO_2 gas injection.

CONCLUSIONS

Comparing water injection and miscible water alternating CO_2 gas injection, scenarios show that miscible water alternating CO_2 gas injection scenario has more efficiency comparing to water injection and produced oil and recovery factor would be greater in this method.

- Comparing water injection and miscible water alternating CO_2 gas injection, scenarios show that produced water during water miscible water alternating CO_2 gas injection scenario would be less comparing to water injection method.

- Water saturation during water injection scenario is more homogeneous all over the reservoir comparing to miscible water alternating CO_2 gas injection scenario. This would mean that sweep efficiency increases after miscible water alternating CO_2 gas injection.

- Most of Iranian oil reservoirs are fractured. So described method of miscible water alternate CO_2 injection is a suitable and recommended choice to be used in Iranian oil reservoirs.

REFERENCES

1. Thompson, J.L. and Mungan, N. (1968) A Laboratory Study of Gravity Drainage in Fractured Systems under Miscible Condition. The 43rd SPE Annual Fall Meeting, Houston.

2. Christensen, J.R., Stenby, E.H. and Skauge, A. (2001) Review of WAG Field Experience. SPE Reservoir Evaluation and Engineering, 4, 97. http://dx.doi.org/10.2118/71203-PA

3. Kulkarni, M.M. and Rao, D.N. (2005) Experimental Investigation of Miscible and Immiscible Water-Alternating-Gas (WAG) Process Performance. The Craft and Hawkins Department of Petroleum Engineering, Louisiana State University, Baton Rouge.

4. Dorostkar Mohammad, J. and Mohebi, A. (2009) Laboratory Evaluation of Warm Water Alternating Immiscible Gas Injection Warm of Carbon Dioxide for Enhanced Oil Recovery in Fractured Model. Enhanced Oil Recovery—Iranian Chemical Engineering Journal, 8.

5. Taghavi, S.A. and Moradi, A. (2009) Compared Using Simulated Oil Recovery Factor of Miscible Gas Injection and Immisible Gas Injection in Iranian Fractured Reservoirs. Enhanced Oil Recovery—Iranian Chemical Engineering Journal, 8.

Chapter 3

Fully Compositional and Thermal Reservoir Simulation

Rustem Zaydullin[a], Denis V. Voskov[a], Scott C. James[b], Heath Henley[c], and Angelo Lucia[d]

[a]Department of Energy Resources Engineering, Stanford University, Stanford, CA, United States

[b]Exponent, Inc., 320 Goddard, Suite 200, Irvine, CA 92618, United States

[c]Department of Chemical Engineering, University of Rhode Island, Kingston, RI 02881, United States

[d]FlashPoint LLC, Narragansett, RI 02882, United States

ABSTRACT

Fully compositional and thermal reservoir simulation capabilities are important in oil exploration and production. There are significant resources in existing wells and in heavy oil, oil sands, and deep-water reservoirs. This article has two main goals: (1) to clearly identify chemical engineering sub-problems within reservoir simulation that the PSE community can potentially make contributions to and (2) to describe a new computational framework for fully compositional and thermal reservoir simulation based on a combination of the Automatic Differentiation-General Purpose Research Simulator (AD-GPRS) and the multiphase equilibrium flash library (GFLASH). Numerical results for several chemical engineering sub-problems and reservoir simulations for two EOR applications are presented. Reservoir simulation results clearly show that the Solvent Thermal Resources Innovation Process (STRIP) outperforms conventional steam injection using two important metrics – sweep efficiency and oil recovery.

INTRODUCTION

Modeling and simulation to predict long-term performance of oil recovery methods (i.e., reservoir simulation) is a topic studied for over 50 years (see Douglas et al., 1959, Price and Coats, 1974 and Todd et al., 1972). Early reservoir models (e.g., black-oil reservoir models) were typically based upon rigorous mass-balance equations for key species (oil, water, and gas), but only used approximate phase equilibria (e.g., no oil dissolved in the water phase) and/or neglected energy balances. By the 80's, reservoir simulation had reached a level of maturity to warrant the first Society of Petroleum Engineers (SPE) Comparative Solutions Project on 3-D Black Oil Reservoir Simulation (Odeh, 1981) in which seven different companies participated. To date, there have been ten separate comparative solution projects sponsored by the SPE with topics that include three-phase behavior, steam injection, horizontal wells, and effective grid-generation and up-scaling techniques. These Comparative Solutions Projects papers are useful for readers new to reservoir simulation or those simply interested in learning more about challenging issues in this area.

Today, reservoir simulation has reached a point where advanced concepts such as dual-porosity models, rigorous phase behavior, energy-balance considerations, fully implicit time stepping with Newton's method to solve the reservoir model equations at each time step, iterative linear solvers, finite difference, and/or analytical Jacobian matrices (to name a few) are available as modeling components.

There remains considerable oil in place (OIP) in many reservoirs that are either in current operation or have been shutdown (often with infrastructure remaining in place). There are also large amounts of fossil fuels in heavy oil, oil sands, and deep-sea reservoirs – but these hydrocarbons are more challenging and more costly to produce. An increase in production at a standard oil field of just 1% can represent a \$25 B opportunity. Many oil producers are considering enhanced oil recovery (EOR) methods such as steam injection and in situ CO_2 + steam injection (i.e., Solvent Thermal Resource innovative Process or STRIP) as a means of increasing recovery. Modeling STRIP, and other advanced EOR methods necessarily requires both fully compositional and thermal reservoir flow simulation capabilities, something that remains challenging.

Perhaps it is not surprising that various aspects (or sub-problems) of fully compositional and thermal reservoir modeling and simulation are, in many ways, similar to modeling and equation-solving task associated with the kinds of chemical processes with which the process systems engineering (PSE) community and readership of Computers & Chemical Engineering are familiar. These sub-problems include

- Multi-phase equilibrium or flash.
- Chemical reaction equilibrium.
- Combined chemical and phase equilibrium.
- Adiabatic flame temperature determination.
- Heat and mass transfer in porous media.
- Models consisting of differential algebraic equations (DAEs).
- Nonlinear equation-solving using Newton and trust region methods.
- Iterative linear equations solving.

Thus it is our opinion that chemical engineers, particularly those in the process systems engineering (PSE) community, are in a unique position to make significant contributions to various aspects of reservoir simulation.

In this paper, we present an advanced reservoir modeling and simulation framework for fully compositional and thermal reservoir simulation and subsequently apply this simulation framework to a comparative study of steam injection and STRIP in EOR applications. We also identify those sub-problems to which the PSE community can contribute. Accordingly, this work is organized as follows. Section 2 gives an overview of the relevant literature. In Section 3, a generalized reservoir model is presented; it includes model equations for both the reservoir and the bulk-phase length scales. Coupling between the reservoir and other constitutive equations needed to close the model (e.g., multi-phase equilibrium flash, viscosity correlations, Darcy's law, heat conduction, etc.) are also described. In Section 4, details that describe how model equations are formulated and solved at various computational levels are given. Specific algorithmic features of the coupled methodology are also presented. In Section 5, steam injection and STRIP are introduced along with common metrics used to evaluate thermal EOR techniques. In Section 6, several relevant sub-problems are presented and solved prior to the application of this new reservoir simulation framework to two reservoir examples that demonstrate modeling and simulation capabilities and quantify the reliability and computational efficiency of the approach. A quantitative comparison of steam injection and STRIP is provided for the first reservoir simulation example using common performance metrics. The second example compares the performance of a modified compositional space adaptive tabulation (CSAT) with the conventional multiphase flash approach. Finally, in Section 7 conclusions of this work are drawn and future needs are highlighted while in Section 8 some additional sub-problems of interest to the PSE community are identified.

LITERATURE SURVEY

The focus of this article is numerical reservoir simulation, which comprises a vast body of literature and thus it is not possible to survey all relevant scientific papers. Therefore in this section only a summary of those papers and numerical methods directly relevant to the modeling and simultaneous solution of numerical reservoir models is presented. We refer the reader to the book by Peaceman (2000) for an introduction to the fundamentals of reservoir modeling and simulation and a description of some of the foundational numerical methods that have been developed. A secondary focus of this manuscript is to identify sub-problems within a larger reservoir simulation that are clearly within the skill set of process systems engineers.

Some of the earliest work in numerical reservoir simulation dates back to 1959 and the pioneering work ofDouglas et al. (1959), who developed numerical methods for the simultaneous solution of time dependent two-phase flow problems in one and two spatial directions.

Governing partial differential equations (PDEs) describing conservation of mass and flow were converted to nonlinear algebraic equations using difference approximations and the resulting nonlinear algebraic model equations were then solved using various numerical methods including alternating-directions implicit, Jacobi iteration, successive over-relaxation, Gauss-Seidel iteration, and other established techniques. We refer the reader to the book by Ortega and Rheinboldt (1970) for a comprehensive description of these numerical methods. The work of Douglas et al. (1959) was later extended to three spatial dimensions by Coats et al. (1967) and to three-phase flow problems by Peery and Herron (1969) and Sheffield (1969). Other journal articles that address additional physics in reservoir simulations and solve model equations simultaneously include those by Snyder (1969), Settari and Aziz (1974), and Trimble and McDonald (1976). Key differences among many of the early approaches to reservoir simulations reside largely in model formulation and the methods used to solve the resulting algebraic model equations. These differences persist today. Note that all of the topics just described are familiar to the PSE community – dynamical equations describing conservation of mass and energy, differencing, and nonlinear equation solving.

State-of-the-art reservoir simulation has moved to two basic nonlinear formulations – a natural formulation (Coats, 1980) and a molar formulation (Acs, 1985). Large sets or subsets of nonlinear algebraic equations result from discrete representations of the governing PDEs that describe the spatial and temporal evolution of the system. The most commonly used approaches for discrete representation are finite-difference or finite-volume approximations on structured or unstructured grids. The resulting algebraic equations are generally solved simultaneously using variants of Newton's method, although various forms of model reductions are also used. In the natural formulation, pressure, temperature, saturation, and all phase compositions for all grid blocks comprise the set of unknown variables. In the molar formulation, which is probably the formulation that is more familiar to engineers in the PSE community, pressure, temperature, and overall compositions (or total component mass) are the unknown variables. While there are many approaches to model formulation and solution, some of the more commonly used methods are differentiated by the temporal discretization scheme – Fully Implicit Method (FIM), IMplicit Pressure Explicit Saturation (IMPES), IMplicit Pressure and SATuration (IMPSAT), and Adaptive Implicit Methods (AIM).

For the description of different solution techniques that follows, we use the natural formulation in a reservoir application in which the pressures, saturations, and phase compositions for all grid blocks are the unknown variables. In the FIM, all pressures, saturations, and compositions of all phases are computed simultaneously at each time step. One of the key advantages of the FIM is that it is unconditionally stable. In contrast, the IMPES methodology treats all terms that depend on saturation and compositions, except the transient terms, as explicit functions of these variables. This allows saturation and composition to be decoupled from the pressure, resulting in a smaller subset of equations to be solved simultaneously, which reduces overall computational demand. However, because IMPES involves some explicit terms, integration may not be numerically stable in regions where volumetric flows are large. As a result, the computational time saved by reducing the size of the system of nonlinear equations can often be negated by smaller time stepping and, in the worst case, can lead to model failure. IMPSAT is similar IMPES, except that IMPSAT treats pressures and saturation variables for all grid blocks implicitly and phase compositions for all grid blocks explicitly. AIM, on the other hand, is intended to marry the best characteristics of FIM, IMPSAT, and IMPES by switching between different solution methods using one or more prescribed metrics, as solution stability demands. For example, AIM might use the spectral radius of a transformation matrix in the residuals of the mass conservation equations to decide when to switch from FIM in regions where instabilities in IMPES are likely, but use IMPES everywhere else (Cao, 2002). A good survey of the numerical characteristics of FIM, IMPES, and AIM is given by Marcondes et al. (2009). Regardless of the formulation, many current solution methods use some form of iterative linear equation solver (e.g., GMRES or other Krylov subspace methods) with pre-conditioning to solve the linear system of equations that determines the Newton correction to the variables at each time step.

RESERVOIR MODEL EQUATIONS

The equations describing the time evolution of fluid composition, temperature, and pressure in a reservoir comprise a set of coupled, nonlinear PDEs that describe conservation of mass, energy, and momentum. In addition, various thermo-physical properties, equilibrium (or non-equilibrium) behavior of fluid phases, properties of porous media, and well-configuration specifications are included as algebraic constraints to the governing PDEs. In this article, the governing PDEs are represented in discrete form using finite-volume discretization and, when used with additional constraints, they form a large set of nonlinear algebraic equations. In this section, the reservoir equations as well as other constitutive equations are described.

Reservoir Model Equations in General Form

The nonlinear time-dependent PDEs that represent conservation of mass and energy in a reservoir are given by

$$\frac{\partial}{\partial t}\left(\phi\sum_{k=1}^{P}\rho^k x_i^k S^k\right) - \sum_{k=1}^{P}\left(\rho^k x_i^k V^k + S^k J_i^k\right) - Q_i = 0, \quad i = 1, \ldots, C \tag{1}$$

and

$$\frac{\partial}{\partial t}\left[(1-\phi)\rho_M U_M + \phi\sum_{k=1}^{P}\rho^k U^k S^k\right] - \sum_{k=1}^{P}\left(\rho^k H^k V^k + S^k G^k\right) - Q_E = 0 \tag{2}$$

where ϕ is the porosity of the porous media, ρ denotes molar density, x is composition in mole fraction, S is saturation, V is volumetric flow, J is molar diffusion flux, which is usually ignored for large scale applications, and Q is a source or sink term. In Eq. (2), U denotes internal energy, H is enthalpy, and G is heat conduction flux. The subscript i denotes a given component while the superscript k denotes a given phase. Summations are over all phases $k = 1, \ldots, P$. C is the total number of components in the mixture. The subscript M in Eq. (2) denotes the rock media whereas the symbol $\triangledown$ denotes the gradient of a vector.

Phase Equilibrium in General Form

Phase equilibrium in a finite volume cell (or grid block) is described by the equality of partial fugacities for all components in all phases, clearly a topic on which publications in the chemical engineering literature abound. In particular,

$$f_i^1 = f_i^2 = \cdots = f_i^k, \quad i = 1, \ldots, C; \quad k = 1, \ldots, P$$

(3)

where f_i^k denotes the partial fugacity of component i in phase k and is given by

$$f_i^k = x_i^k \varphi_i^k p$$

(4)

where ϕ is the fugacity coefficient of component i in phase k and p is pressure.

Conservation of mass within any grid block is represented by a set of component mass-balance equations

$$\rho_T z_i - \sum \rho^k S^k x_i^k = 0, \quad i = 1, \ldots, C$$

(5)

where ρ_T and z_i are the total density and mole fraction of component i in the cell. Note that there is some overlap in symbols because standard notation in reservoir engineering and chemical engineering thermodynamics each use the same symbol to denote different quantities. We caution the reader to pay careful attention to context so the meaning of a symbol is clear.

Finally, in the natural formulation, Eqs. (1), (2) and (3) are solved simultaneously and do not require a separate solution of flash problem. For the molar formulation, overall composition, temperature, and pressure of a given finite volume are specified, and, as a consequence, Eqs. (3), (4) and (5), which constitute the classical isothermal, isobaric (Tp) flash problem, must be solved separately for the number and type of equilibrium phases and their corresponding compositions and densities.

Equation of State

The topic of equations of state (EOS) is intimately familiar to the PSE and thermodynamics communities of chemical engineering and, in general, EOS are required to model reservoir fluids. This is because some of the components (e.g., CH_4, N_2, CO_2) and/or mixtures of components can be supercritical at various conditions of temperature and pressure in a reservoir. Using an EOS, all phase properties (i.e., density, fugacity coefficients, fugacities, chemical potentials, enthalpies, etc.) can be readily computed. Furthermore, cubic equations are preferred over more complex equations like Statistical Associating Fluid Theory because they have a lower computational overhead and provide results that are within acceptable accuracy. As described later in this article, GFLASH allows the user to select from a number of more commonly used cubic EOS.

Other Constitutive Equations

Other constitutive equations are also needed to close the numerical model and allow proper integration of Eqs. (1) and (2). These constitutive equations include Darcy's Law, heat conduction, and when relevant, diffusion equations – again all topics familiar to the PSE community in traditional applications such as heat and mass transfer in catalyst pellets, non-equilibrium models in multi-stage distillation, as well as more recent applications in bio-medical modeling and simulation of the brain among others.

Darcy's Law

Darcy's law describes the volumetric flow of each phase, V^k, through porous media as

$$V^k = -\left(\frac{\kappa \hat{R}_i^k \rho^k x_i^k}{\mu^k}\right) \nabla(p + \rho^k x_i^k gz)$$

(6)

Where k is an intrinsic rock or soil permeability, $\hat{R}$ is relative permeability, μ is viscosity, g is the acceleration due to gravity, and z is the coordinate in the direction of gravity.

Heat Conduction Equations

The heat conduction equation is

$$G^k = -K^k \nabla T$$

(7)

where K is the thermal conductivity and T is absolute temperature.

Equation Coupling

The conservation of mass and energy, flow, and conduction through porous media described by Eqs. (1),(2), (3), (4), (5), (6) and (7), and the equations describing the conservation of mass, conservation of energy with heat losses to the surroundings, and phase equilibrium form a large system of strongly coupled nonlinear algebraic equations. In a hierarchical sense, the EOS lies at the innermost level of the computations and provides the phase densities. Phase densities are used to calculate fugacity coefficients, and fugacities to determine the type and amounts of each phase present in a grid block (i.e., by solving the traditional chemical engineering Tp flash). GFLASH calculated phase densities and composition are then used to determine the unknown variables at the reservoir level (e.g., pressures, saturations, and temperatures) as well as heat conduction fluxes, and the flow of phases through the porous media.

IMPLEMENTATION

As noted in the literature survey, several computer implementations and methods of solving the model equations described in Section 3 are available. In this subsection, we describe the specific implementation of the reservoir model, well models, and constitutive equations associated with heat conduction. The reservoir modeling software is called Automatic Differentiation – General Purpose Research Simulator (AD-GPRS). AD-GPRS was originally developed and is currently maintained by the SUPRI-B group in the Energy Resources Engineering Department at Stanford University. It enjoys widespread use throughout the reservoir and petroleum engineering communities. AD-GPRS is written in C++. The EOS and flash calculations are implemented in a suite of FORTRAN programs called GFLASH, which was developed and is maintained by A. Lucia, and may be of particular relevance to the PSE community.

AD-GPRS

AD-GPRS is an advanced reservoir simulator with wide ranging capabilities that include

- flexible treatment of all nonlinear physics,
- a fully thermal-compositional formulation for any number of phases,
- multi-phase CSAT for efficient and robust computation of phase behavior,
- a variety of discretization schemes in time and space,
- thermal geo-mechanical modeling including the effects of fractures,
- a fully coupled, thermal, multi-segmented well model with drift-flux, and
- an adjoint-based optimization module.

There are, of course, many details associated with AD-GPRS (Voskov & Zhou, 2012); its main features are summarized here.

Formulations

Both natural and molar formulations are available in AD-GPRS (Voskov & Tchelepi, 2012). Regardless of formulation, the primary dynamic model equations describing the time evolution of material and energy in a reservoir given by Eqs. (1) and (2) are appended with a number of constraint equations to form a differential algebraic equation (DAE) system. The algebraic constraint equations include

- fugacity constraints [i.e., Eq. (3)].
- summation equations for the mole fractions in each phase.

$$1 - \sum_{i=1}^{C} x_i^k = 0,$$

(8)

- saturation summation equations

$$1 - \sum_{k=1}^{P} S^k = 0,$$

(9)

- volume balance constraints when total mass variables are used

$$\phi \rho_T V - \sum_{i=1}^{C} n_i = 0,$$

(10)

where n_i is the overall number of moles of component i in a grid block (fixed for each GFLASH calculation) and v is the volume of a grid block.

Eqs. (1), (2), (3), (6), (7), (8) and (9), and (10) comprise a DAE representation of the reservoir equations.

Discretization

The DAE system described by Eqs. (1), (2), (3), (6), (7), (8) and (9), and (10), is converted into a set of nonlinear algebraic equations using finite volume spatial and temporal discretizations. Note that Eqs. (1),(2), (3), (6) and (7), and (8) are essentially the same equations used to model traditional steady and unsteady-state chemical processes.

Spatial Discretization

Spatial representation of a reservoir in discrete form in AD-GPRS uses the Multi-Point Flux Approximation to account for the geometry of fluxes across interfaces (see Zhou et al., 2011 for details). Consider the flux across the interface shared by two cells, denoted by j and j_1, and assume that the normal vector at the interface has an orientation that points into cell j. The overall flux of component i from j_1 to j is given by

$$F^{j,j_1} = \sum_{k=1}^{P} x_i^k \rho^k \lambda^k (\Phi^{j,j_1})^k$$

(11)

where k is the mobility of phase k. Here all quantities except $k(j,j_1)$ are taken in upstream of flow direction. $k(j,j_1)$ is called the geometric part of the flux of phase k and is approximated by

$$(\Phi^{j,j_1})^k = \sum_{k=1}^{P} \theta^{j,j_1} \left[p_j^k - p_{j_1}^k + g(\gamma^{j,j_1})^k d^j \right]$$

(12)

where the summation in Eq. (12) is over the number of data points associated with the flux across interface $\{j, j_1\}$ (only one for a Two-Point Flux Approximation), $^{j,j1} > 0$ is the transmissibility coefficient average on interface $\{j, j_1\}$, dj is the depth of cell j, and $k(j,j1)$ is the mass density of phase k averaged at the interface $\{j, j_1\}$.

Similarly heat (energy) flux can be expressed as

$$E^{j,j_1} = \sum_{k=1}^{P} [\rho^k H^k \lambda^k (\Phi^{j,j_1})^k + S^k K^k \theta_g^{j,j_1} (T_j - T_{j_1})]$$

(13)

where $\theta^{j,j1}$ is the geometrical part of the transmissibility coefficient assuming Two-Point Flux Approximation (TPFA) for the conduction term (Voskov and Zhou, 2012).

Temporal Discretization

Temporal discretization by implicit integration is unconditionally stable. AD-GPRS has a number of the commonly used temporal discretizations – FIM, IMPES, IMPSAT, and AIM. As noted in Section 2, each of these methods represents a different approach where different unknown variables and equations are treated either explicitly or implicitly. In AD-GPRS, FIM, IMPES, and IMPSAT are all considered special cases of AIM. Finally, Courant–Friedrich–Lewy (CFL) stability criteria are used to adaptively determine the level of implicitness to solve the model equations.

Solution of Nonlinear Algebraic Equations

After assembling the Jacobian matrix, the Newton–Raphson method solves the linear system of equations at each iteration. The relations, other than the mass conservation equations, are treated as constraints that are local to a grid block. To minimize the size of the global linear system, a Schur-complement procedure is applied to the full Jacobian matrix of each block to express the primary (mass conservation) equations as a function of the primary variables only (Cao, 2002). After the size of the system is reduced, the resulting global linear system of equations is solved for the primary variables. An iterative linear equation solver with pre-conditioning is used to solve the linear system.

After the linear system is solved, the computed changes to the primary variables are used with the secondary equations to determine changes in the secondary variables locally in each grid block. Next, the nonlinear variables are updated using different strategies and safeguards to ensure that the solution remains within physical boundaries. Convergence of Newton–Raphson iteration depends on aspects that include (1) any corrections to updated variables that employ safeguards, (2) various chopping strategies for different unknowns, and (3) the choice of time step. Several strategies for updating variables and time-step choice are implemented in AD-GPRS (Voskov and Zhou, 2012).

Phase Behavior Computations

In this section, different approaches to phase behavior computations in AD-GPRS are described including the use of intermittent flash solutions and CSAT.

Intermittent Flash Problem Solutions

For phase behavior computations, AD-GPRS uses a two-stage procedure. In the first stage, the number of phases that exist in each grid block is determined. This can be obtained using Gibbs energy minimization or phase stability analysis (Michelsen, 1982a). In the second stage, flash calculations are performed to determine the compositions of the existing phases (Michelsen, 1982b). At both stages a combination of Successive Substitution Iteration and Newton's method is used.

As an alternative to this two-stage strategy, a generalization of the negative-flash based approach ofWhitson and Michelsen (1989) can be used (Iranshahr et al., 2010). Here it is assumed that the number of phases present is the maximum possible, and then Eqs. (3) and (5) are solved, allowing for phase fraction to be less than zero, or greater than one. When the phase fractions of a converged negative flash procedure are negative, fewer existing phases are assumed and a similar procedure for this reduced system may be required (Iranshahr et al., 2010).

Compositional Space Adaptive Tabulation (CSAT)

Solving flash problems for all grid blocks over all nonlinear iterations and time steps is computational demanding. To improve the performance of phase behavior computations in reservoir simulation, the CSAT approach originally developed by Voskov and Tchelepi, 2009a and Voskov and Tchelepi, 2009b is used. CSAT adaptively stores a discrete set of tie-lines at different pressures and temperatures to represent phase behavior during reservoir simulation. This collection of tie-lines is interpolated and used to look up the phase state of the mixture at a particular pressure and temperature. In addition, the number of tie lines is collected adaptively based on the specific attributes of a compositional solution during a reservoir simulation.

CSAT completely replaces the need for phase stability tests and provides good initial guesses for the standard Tp flash computations.

Compositional Space Parameterization (CSP)

The compositional space parameterization (CSP) method (Voskov and Tchelepi, 2009a and Zaydullin et al., 2013) is based on casting the nonlinear governing equations (1) and (2), including thermodynamic phase equilibrium constraints (3), in terms of the tie-simplex ( ) space. During a simulation, the   space is adaptively discretized using supporting tie-lines. The coefficients for the governing system of equations, including the phase compositions, densities, and mobilities, are computed using multi-linear interpolation in the discretized space.

Using the CSP methodology, phase behavior computations can be replaced by an iteration-free look-up table procedure during the course of a reservoir simulation, removing the need for standard EOS computations (phase stability and flash). Also, it is important to note that the error associated with multi-linear interpolation is bounded and decreases with grid (or space) refinement (Zaydullin et al., 2013) and therefore only a limited number of supporting tie-lines are needed for the accurate representation of phase behavior. That, in turn, leads to significant gains in computational efficiency.

Tie-lines or tie-simplexes needed for CSAT and CSP can be parameterized using the generalized negative flash procedure (Iranshahr et al., 2010) or with GFLASH (Section 4.2).

GFLASH

GFLASH is a FORTRAN suite that models and solves the traditional chemical engineering multi-phase, multi-component isothermal, isobaric (Tp) flash problem. That is, given an overall composition for a fluid mixture, a temperature, and pressure, GFLASH determines the number of phases that exist at equilibrium and their corresponding compositions, fugacities, densities, and enthalpies. In this section, formulations, overall solution strategies, and methods of solution are described.

Equations of State

A number of the commonly used cubic EOS with and without volume translation are implemented in GFLASH. The EOS available include the

- Soave–Redlich–Kwong (SRK) equation (Soave, 1972),
- SRK with the Peneloux volume translation (SRK+) equation (Peneloux et al., 1982),
- Predictive SRK (PSRK) equation (Holderbaum & Gmehling, 1991),
- Electrolyte PSRK (ePSRK) equation (Kiepe et al., 2004),
- Peng–Robinson (PR) equation (Peng & Robinson, 1976),

- volume translated PR (VTPR) equation (Ahlers and Gmehling, 2001 and Ahlers and Gmehling, 2002), and
- multi-scale Gibbs–Helmholtz Constrained (GHC) equation (Lucia et al., 2012).

Formulation and Solution

All EOS are formulated as cubic polynomials in compressibility factor, z, in the complex plane in the form

$$f(z) = c_1 z^3 + c_2 z^2 + c_3 z + c_4 = 0$$

(14)

The resulting single variable function, $f(z)$, is solved using Newton's method in the complex plane to find any root to an accuracy of $|f(z)| \leq 10^{-12}$. The cubic polynomial is then deflated to a quadratic equation, which is solved using the quadratic formula to determine the other two roots. This approach removes the need to use an accurate initial guess for Newton's method, guarantees that all three roots will always be found, and is actually faster than using the analytical solution to a cubic polynomial.

Root Assignment

Correctly determining which root is liquid and which root is vapor is as important, if not more important, than computing roots to EOS and is particularly challenging under harsh conditions (i.e., high T and high p). The current approach used to assign roots in GFLASH is as follows: For a set of roots given by $\{z_1, z_2, z_3\}$, where any root has the complex variable form $z_k = a_k \pm b_k$, $k = 1, 2, 3$, we define

$$Z^L = \min(|Z_l|) \quad \text{and} \quad Z^V = \max(|Z_l|)$$

(15)

where $|z_k|$ denotes the complex absolute value function given by $|Z_i| = \sqrt{a_{i2} + b_{i2}}$ and the superscripts L and V denote liquid and vapor, respectively. Phase densities are easily computed from the compressibility factors, zL and zV, using the expression

$$\rho = \frac{p}{z^k RT}$$

(16)

Flash Problem Formulations and Method of Solution

The flash problem is really two problems – a phase stability problem and a phase equilibrium problem. In GFLASH, the formulations of the phase stability and phase equilibrium conditions use the dimensionless Gibbs free energy of mixing, G/RT, and the dimensionless Gibbs free energy, G/RT, respectively.

Phase Stability

Minima in G/RT often turn out to be inexpensive and good approximations for points of tangency. The necessary conditions for a minimum in G/RT are formulated in terms of the equality of dimensionless chemical potentials, μi, $i = 1, ..., C$. For the phase split (or phase stability) problem, which is always a two-phase determination, the model equations are given by

$$F(x) = \frac{\left[(\mu_1 - \mu_i^0) - (\mu_C - \mu_C^0)\right]}{RT}, \quad i = 1, ..., C - 1$$

(17)

where the superscript 0 denotes standard state and the unknown variables in Eq. (17) are the mole fractions, xi, $i = 1, ..., C - 1$, of a single hypothetical phase. Note that this formulation of the phase split problem results from the projection of the dimensionless Gibbs free energy of mixing onto the summation equation [i.e., Eq. (8)].

Phase Equilibrium

Phase equilibrium equations are also formulated in terms of dimensionless chemical potentials using projection onto the conservation of mass equations. Conservation of mass for the phase equilibrium problem is given by

$$n_i - \sum_i n_i^k = 0, i = 1, \ldots, C$$

(18)

where n_i is the overall moles of component i in the system and is fixed, n_i^k is the number of moles of ith component in the kth phase, and the summation in Eq. (18) is over all phases. Note that the phase equilibrium problem is formulated in terms of mole numbers, not mole fractions, because it is a way of exploiting many of the useful mathematical properties of partial molar quantities.

Phase equilibrium is defined by the equality of dimensionless chemical potentials given by

$$\frac{\mu_i^1}{RT} = \frac{\mu_i^2}{RT} = \cdots = \frac{\mu_i^k}{RT}, \quad i = 1, \ldots, C; \quad k = 1, \ldots, P$$

(19)

for any number of total phases, P. Eq. (19) is expressed in the form

$$F(n_i^1, n_i^2, \ldots, n_C^P) = \left(\frac{\mu_i^1}{RT} - \frac{\mu_i^k}{RT} \right) = 0, \quad i = 1, \ldots, C; \quad k = 1, \ldots, P$$

(20)

and then projected onto the mass balance constraints in Eq. (18) to reduce the size of the phase equilibrium problem and to ensure that conservation of mass is satisfied at each iteration.

Other Modeling Capabilities in GFLASH

GFLASH also has the capability of solving chemical reaction equilibrium problems and combined chemical and phase equilibrium problems, topics that are both familiar to the PSE community and important in various applications of EOR. For example, in applications of STRIP, partial oxidation of methane is used to generate in situ CO_2 and steam. In other EOR applications, where production water is re-used in order to defray the high cost of purchasing municipal water or expensive water treatment, salt precipitation in the presence of multiple fluid phases, which is a combined chemical and phase equilibrium problem, can be a serious operational problem. These sub-problems are also solved by Gibbs free energy minimization within the GFLASH framework.

Method of Solution

GFLASH uses a trust region method to solve both the phase stability and phase equilibrium model equations. This methodology is a simple version of the terrain methodology developed by Lucia and Feng (2003). When applied to phase stability and phase equilibrium, we restrict the terrain method to look for only one stationary point in each of G/RT and G/RT respectively.

In addition, when solving flash problems, GFLASH alternates between phase stability and phase equilibrium sub-problems, maintaining a monotonically decreasing sequence of values of G/RT until a global minimum identifying the number and type of phases as well as their associated mole numbers is found. Phase stability problems [i.e., Eq. (17)] are solved to an accuracy of $||F(x)||_2 \leq 10^{-6}$, where $||\cdot||_2$ denotes the 2-norm or Euclidean norm. In contrast, phase equilibrium problems (Eq. (20)) are solved to an accuracy of $||F(n)||_2 \leq 10^{-4}$ for two-phase equilibria and 10^{-5} for three-phase equilibria.

The Connection between CSAT and GFLASH

In this section, we describe the connection between rigorous phase stability and flash computations using GFLASH and CSAT. We also describe the interface between AD-GPRS and GFLASH.

Conventional Phase Behavior Computations

The number and types of phases (or phase state) of a mixture in a given grid block can vary. For example, for mixtures that exhibit three-phase behavior, there are seven different possible phase states – three single phase states (i.e., water-rich liquid, vapor, or oil-rich liquid), three different two-phase states (i.e., LLE, water-rich VLE, or oil-rich VLE), or vapor-liquid-liquid equilibrium (VLLE). Thus, the phase state as well as all corresponding phase compositions need to be determined for every grid block on each Newton iteration. For the natural formulation, a three-step procedure is usually used for these computations:

- For any grid block, the current phase state is determined using a phase stability test.

- If the current phase state is different from one on a previous Newton iteration, flash computations are performed to obtain phase compositions.
- Phase properties (i.e., fugacities, densities, enthalpies, etc.) are obtained using known phase compositions.

Because of the complexity of the ADGPRS-GFLASH interface, both a phase stability test and flash computations are performed simultaneously.

Phase Behavior Computations with CSAT

As noted, CSAT can significantly improve the time required for phase behavior computations in fully compositional reservoir simulation (Voskov and Tchelepi, 2009a and Voskov and Tchelepi, 2009b). The general multiphase implementation of CSAT (Iranshahr et al., 2010 and Voskov and Tchelepi, 2009b) is a two-step procedure:

- Computation of supporting tie-simplexes (i.e., tie-triangles for three-phase systems).
- Parameterization of tie-simplex subspace (tie-triangle planes for three-phase systems).

In the original CSAT implementation of Iranshahr et al. (2010), a generalization of the negative-flash idea (Whitson & Michelsen, 1989) for Step 1 and geometrical parameterization (i.e., tracking tie-lines from each side of a tie-triangle) for Step 2 was used. While this approach proves to be robust for challenging three-phase systems, it requires some preliminary knowledge of a multiphase mixture under investigation because the geometry of tie-simplex subspace can be quite complicated.

In this work, we have used a different strategy. First, we use GFLASH to provide fugacities for given pressure, temperature, and phase compositions and the generalized negative flash approach (Iranshahr et al., 2010) is used to find a supporting tie-simplex for the CSAT procedure. Next, an extension of the tie-simplex is adaptively discretized and GFLASH determines the phase state of a model cell. Finally, the collection of tie-simplexes and their extensions are interpolated for a particular pressure and temperature and used to look up the phase state of the mixture.

AD-GPRS/GFLASH Interface

Because AD-GPRS is written in C++ and GFLASH is a FORTRAN suite, the proposed modeling and simulation framework is necessarily mixed language and therefore an interface is needed to communicate information between the two programs.

THERMAL EOR METHODOLOGIES

In this section, steam injection and STRIP are introduced along with common performance metrics used to evaluate thermal EOR methods. We refer the reader to the work of Aziz et al. (1987), which is the 4th SPE Comparative Solution Project: Comparison of Steam Injection Simulators, for an introduction to steam injection and associated simulation challenges.

Steam Injection

Steam injection is generally implemented using surface facilities to generate superheated steam, which is injected into a reservoir through a well. The entering steam heats the formation and lowers oil viscosity, which allows the oil to flow more easily to production wells. In all steam injection methods, surface generation of steam suffers from a number of disadvantages, not the least of which is energy losses (up to 50%) in the piping system and injection well.

Solvent Thermal Resource Innovation Process (STRIP)

STRIP, which has been developed by RII North America, is an environmentally friendly approach to EOR, which is deployed into existing wells, so there is little or no disruption of land. Unlike other steam injection processes, STRIP generates steam and CO_2 by in situ combustion of methane in oxygen, which eliminates energy losses to the injection well and delivers steam directly to the formation. STRIP also provides a co-solvent, CO_2, which enhances oil recovery by swelling oil and lowering viscosity. The STRIP burner can be placed in a number of configurations, but in this work the STRIP burner resides in a vertical section of the injection well. Because the combustion temperature can approach 3000 °C, the STRIP burner is typically cooled using production water, significantly reducing and often removing the need for municipal water. The nominal composition of lumped gases entering a reservoir formation is around 10 mol%, with roughly 6.7 mol% being CO_2.

Performance Metrics

Several common metrics are used to evaluate the performance of a thermal EOR methodology. These metrics include (1) sweep and (2) oil recovery, which, of course, is of primary interest.

NUMERICAL EXAMPLES

In this section, two reservoir simulation examples are presented to elucidate key points, to compare the performance of steam injection and STRIP, and to demonstrate the reliability and computational efficiency of the numerical tools in GFLASH and AD-GPRS. However, prior to presenting results for reservoir simulation with STRIP, a number of traditional chemical engineering sub-problems needed to be solved, including a chemical equilibrium problem, an adiabatic flame temperature problem, and a salt precipitation problem, to clarify and quantify various aspects of the reservoir simulations. All reservoir simulation runs were performed using an Intel(R) Core(TM) 2 Duo CPU E6750 @2.66 GHz, 1.95 GB of RAM.

Example 1: Chemical Equilibrium of STRIP Combustion

As noted, STRIP generates in situ CO_2 and steam by partial oxidation of methane. In a typical application, the reactants are fuel-rich and thus there are a number of 'major' syngas by-products such as H_2 and CO, and un-reacted methane and O_2 in addition to the CO_2 and steam. However, the composition of the combustion product stream is a function of both the O_2/CH_4 ratio and the reaction temperature, the latter of which we do not know. The O_2/CH_4 ratio, which is denoted by r, is an operational decision based on extensive laboratory experimentation and can vary between at 1.6 and 1.9 depending on the application. Note that the stoichiometric ratio of oxygen/methane for complete combustion is 2 and thus STRIP combustion is fuel-rich and thus will produce some syngas (H_2 and CO).

The governing equations for this single vapor phase chemical equilibrium problem are

$$\min \frac{G}{RT} = \sum_{i=1}^{C} \frac{n_i \mu_i}{RT}$$

(21)

subject to mass balances for hydrogen, oxygen, and carbon

$$\sum_{i=1}^{n} a_{ij} n_i = A_j, \quad j = 1, \ldots, J$$

(22)

where J is the number of atomic species, A_j is the total amount of atom j in the system, and a_{ij} is the number of jth atoms in the ith molecular compound. For this example using a basis of 1 mole of CH_4, the mass balances are

hydrogen: $2nH_2 + 0nO_2 + 0nCO + 4nCH_4 + 0nCO_2 + 2nH_2O = 4$ (23)

oxygen: $0nH_2 + 2nO_2 + 1nCO + 0nCH_4 + 2nCO_2 + 1nH_2O = 2r$ (24)

carbon: $0nH_2 + 0nO_2 + 1nCO + 1nCH_4 + 1nCO_2 + 0nH_2O = r$ (25)

Fig. 1 shows the effect of reaction temperature on the composition of STRIP combustion products for temperatures between 300 and 3100 °C at 20 bar. Note that above about 1000 °C, there is very little change in the composition of the combustion products.

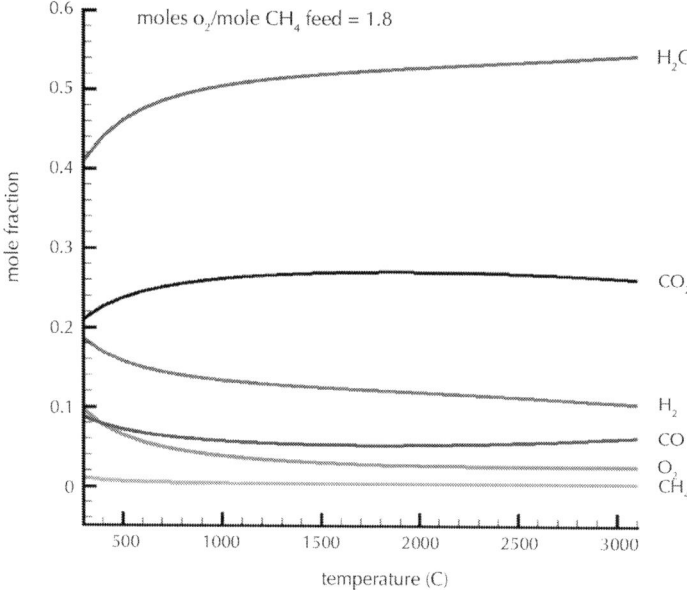

Figure. 1: STRIP combustion product composition vs. temperature for an fuel-rich burner.

Table 1 gives the details of a single chemical equilibrium computation for 2500 °C at 20 bar for an oxygen-to-methane ratio of 1.8. Note that there is a significant amount of 'major' by-product gases, about 20 mol%, and a net production of 0.269696 moles.

Table 1: Numerical results for chemical equilibrium of STRIP combustion

Chemical species	Feed mole fractions	Product gas mole fractions
H_2	0	0.133035
O_2	0.6428571	0.028308
CO	0	0.042660
CH_4	0.3571428	0.002809
CO_2	0	0.280303
H_2O	0	0.512884
Total moles	2.8	3.069696

Example 2: Adiabatic Flame Temperature

While Fig. 1 gives chemical equilibrium results for a wide range of combustion temperatures, the actual temperature of the combustion products for a given set of conditions should be estimated using an adiabatic flame temperature calculation and is coupled to the previous chemical equilibrium problem because the temperature and composition are interdependent. However, because product gas compositions are relatively weak functions of temperature above about 1000 °C, we can effectively decouple the problems and solve the adiabatic flame temperature problem using the product gas compositions shown in Table 1. The corresponding problem formulation, which is rather simple and can be found in several undergraduate level thermodynamics textbooks, is shown in Eq. (26).

$$0 = \Delta H_R^0 + \sum_{i=1}^{C} \int_{T_0}^{T} n_i C_{p,i} \, dT$$

(26)

where ΔH_R^0 is the standard heat of reaction, T_0 is a reference temperature and equal to 25 °C, and Cp,i is the heat capacity for the i th component. Data for ΔH_R^0 and Cpi are given in Appendix B. The calculated adiabatic flame temperature for an oxygen-to-methane ratio of 1.6 is 3343.975 °C.

Example 3: Salt Precipitation

Use of production water in EOR processes can reduce the cost of purchasing municipal water or operating an on-site water treatment plant and concomitantly lower environmental impact. The main challenge associated with the use of production water is the presence of ions and the potential for salt precipitation. To lower the potential for precipitation, production water can be mixed with clean water. In the case of STRIP, production water is mixed with in situ generated steam for two purposes – to generate additional steam and to cool the STRIP combustion burner. Table 2 gives an illustration of the compositions of production water and the combined feed for a STRIP application to a real reservoir in Saskatchewan, Canada before and after mixing. Note that production water analyses are generally reported as molality, whereas mass or mole fractions are generally used in flash calculations.

Table 2: Composition of production water and combined feed for STRIP

Chemical species	Production water molality (mol/kg H_2O)	Mole fractions of combined feed
H_2	0	0.016793
O_2	0	0.003229
CO	0	0.006459
CH_4	0	0.000388
CO_2	0	0.043532
Na^+	1.100299	0.017692
K^+	0.007587	0.000122
Ca^{2+}	0.029479	0.000474
Cl^-	1.166844	0.018762
SO_4^{2-}	0.000023	3.68×10^{-7}
H_2O	0.983278	0.892549
Total		1.000000

The real concern regarding precipitation comes from the fact that the combustion products from STRIP are very hot and thus the amount of liquid available to dissolve ions, even after mixing, might be quite small if too much vaporization occurs. Remember, the main purpose of STRIP is to inject enough steam and CO_2 into the reservoir for improved oil recovery. What this means is that the desired fluid stream entering the reservoir (i.e., at the injection well bore) should have a relatively high vapor fraction, say between 0.7 and 0.8. To determine whether or not salt precipitation will occur, we must therefore solve a combined chemical and phase equilibrium flash problem at high temperature. Salt precipitation is a heterogeneous chemical equilibrium problem and must be determined by comparing equilibrium solubility products, K_{sp}, to ion solubility products, Q_{sp}, for all possible molecular salts as shown in Eq. (27).

$K_{sp}^k > Q_{sp}^k$, then aqueous liquid is under-saturated with molecular salt k $\Big\}$
$K_{sp}^k = Q_{sp}^k$, then aqueous liquid is saturated with molecular salt k $\qquad k = 1, \ldots, n_s$
$K_{sp}^k < Q_{sp}^k$, then the aqueous liquid is super-saturated with salt k $\Big\}$

$$(27)$$

where ns is the number of molecular salts. In this example, there are six possible molecular salts: NaCl, KCl, $CaCl_2$, Na_2SO_4, K_2SO_4 and $CaSO_4$. The standard Gibbs free energy and enthalpy of formation data used to compute K_{sp} is shown in Appendix B. Moreover, it is entirely possible to compute multi-phase equilibrium flash solutions that are supersaturated and meta-stable; thus simultaneously satisfying conditions of multi-phase and chemical equilibrium is quite challenging. Table 3 shows a meta-stable VLE flash solution for the combined feed in Table 2 at 255 °C and 18 bar computed using the GHC equation of state.

Table 3: Meta-stable flash solution to combined chemical and phase equilibrium problem

Quantity	Mole fractions of combined feed	Aqueous liquid	**Vapor**
H_2	0.016793	7.9788×10^{-7}	0.033336
O_2	0.003229	2.8861×10^{-6}	0.006408
CO	0.006459	7.7968×10^{-7}	0.012821
CH_4	0.000388	1.7858×10^{-7}	0.000770
CO_2	0.043532	9.0811×10^{-5}	0.086332
Na^+	0.017692	0.033586	0
K^+	0.000122	0.000247	0
Ca^{2+}	0.000474	0.000961	0
Cl^-	0.018762	0.038027	0
SO_4^{2-}	3.68×10^{-7}	7.4585×10^{-7}	0
NaCl	0	0	0
H_2O	0.892549	0.924811	0.860333
Phase fraction	1.000000	0.496276	0.503724
Density (kg/m³)		831.985	8.464
G/RT	2.53535	2.496677	

Note that the supersaturated VLE solution has a lower value of G/RT than the single phase solution. Table 4 gives the values of K_{sp} and Q_{sp} for all six molecular salts at aqueous liquid phase conditions given in Table 3. Note that Table 4 clearly shows that $K_{sp}^{NaCl} < Q_{sp}^{NaCl}$ and therefore NaCl will precipitate. The molar amount that will precipitate, S_{NaCl}, is easily computed using the following mass balance

$$SNaCl = FxNaCl - _{[c]}Na_+(nH_2OMWH_2O)/1000 \qquad (28)$$

where Fx_{NaCl} is the molar amount of NaCl in the feed, $_{[c]}Na_+$ is the solubility limit of Na^+ in the aqueous liquid, nH_2O is the number of moles of water in the aqueous liquid, and MWH_2O is the molecular weight of water.

Table 4: Equilibrium and ion solubility products for Example 3.

Molecular salt	Ksp	Qsp
NaCl	4.3516	4.9124
KCl	7.0513	0.033875
$CaCl_2$	1.9765	0.30040
Na_2SO_4	2.6759	0.000207
K_2SO_4	7.0259	9.8612×10^{-9}
$CaSO_4$	0.27929	2.5815×10^{-6}

Table 5 gives the global minimum vapor-liquid-solid equilibrium solution for the same combined feed conditions and there are several important points to note regarding this equilibrium solution.

- The VLE + salt solution has a lower dimensionless Gibbs free energy than either the single phase solution or the supersaturated VLE solution.
- The STRIP criterion of 0.7–0.8 vapor fraction has been met in the final solution.
- Salt precipitation is potentially a serious concern in this application of STRIP unless production water is mixed with clean water.

Table 5: Global minimum solution to combined chemical and phase equilibrium problem

Quantity	Mole fractions of combined feed	Aqueous liquid	Vapor	Solid salt
H_2	0.016793	6.8490×10^{-7}	0.023735	0
O_2	0.003229	2.3922×10^{-6}	0.004563	0
CO	0.006459	6.4952×10^{-7}	0.009129	0
CH_4	0.000388	1.4567×10^{-7}	0.000548	0
CO_2	0.043532	7.1298×10^{-5}	0.061500	0
Na^+	0.017692	0.033746	0	0
K^+	0.000122	0.000419	0	0
Ca^{2+}	0.000474	0.001629	0	0
Cl^-	0.018762	0.037426	0	0
SO_4^{2-}	3.68×10^{-7}	1.2654×10^{-6}	0	0
NaCl	0	0	0	1.000000
H_2O	0.892549	0.926701	0.900524	0
Phase fraction	1.000000	0.290190	0.701886	0.007924
Density (kg/m^3)		813.901	8.245	2166.642
G/RT	2.53535	1.80108		

Example 4: Flash Level Reliability Testing

A high level of reliability is needed at the flash level for successful reservoir simulations. Even a single failure in one grid block can cause the entire reservoir simulation to fail. To ensure reliability at the flash level, several phase diagrams similar to the ones shown in Fig. 2 and covering the entire composition space are usually generated using GFLASH for a number of different temperatures and pressures prior to running a reservoir simulation. Typically a composition interval of 0.005 is used for each independent composition. Thus, for a three-component mixture, roughly 20,000 composition points are generated for each temperature and pressure.

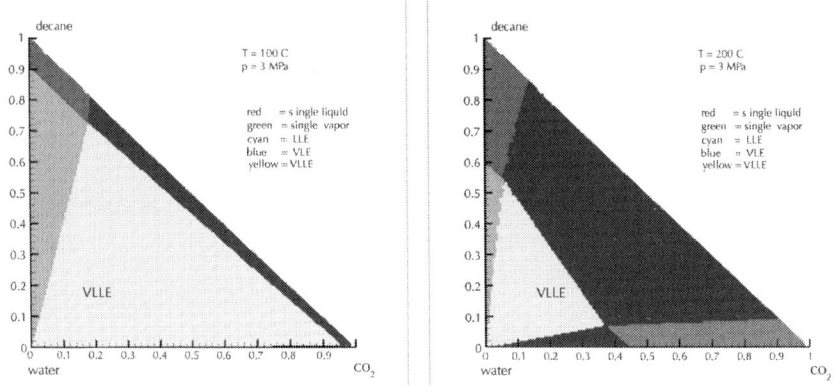

Figure. 2: CO_2–decane–water phase behavior at 100 °C and 200 °C and 30 bars.

This is to ensure that phase boundaries are smooth and that changes in V-only, L-only, VLE, LLE, and VLLE regions make physical sense. Table 6 gives computational details for the rigorous flash tests.

Table 6: Statistics for rigorous flash solutions for CO_2–decane–water at 30 bars using GFLASH.

	T = 373 K	**T = 473 K**
No. of problems	19,532	19,532
No. of liquid-only solutions	687	1559
No. of vapor-only solutions	5	1775
No. of VLE solutions	1876	11,720
No. of LLE solutions	3207	682
No. of VLLE solutions	13,757	3796
No. of function calls	2,114,553	3,255,481
No. of EOS solves	5,448,233	7,709,004
Total solve time (CPU sec)	38.7	26.5

Example 5: Comparison of Steam Injection and STRIP

This first EOR example compares model results for steam injection and STRIP for a 3D heterogeneous reservoir formation containing light oil. Input data are listed in Table 7. In EOR applications it is typical to 'lump' components to reduce computational costs; thus for STRIP all light gases were treated as CO_2, which is the solvent of interest.

Table 7: Input data for Example 5

Quantity	Value
Reservoir dimensions	$360 \times 360 \times 3.6$ m^3
Initial reservoir T and p	290 K, 31 bar
Initial reservoir composition	1 mol% CO_2, 74 mol% oil, 25 mol% water
Porosity, rock heat capacity	0.197% (average), 2.35×10^6 J/m^3 K
Permeability (Upscaled SPE10)[a]	5 orders of magnitude in permeability variations
Injection conditions	Water rate 15 m^3/day
	heat rate 1.3×10^{10} J/day (steam at 518 K)
Injection composition	10% CO_2, 90% water (STRIP) 1% CO_2, 99% water (steam injection)
Production well p	3.45 bar
Time horizon	2000 days

In this example, the model is based on a fragment of the up-scaled SPE10 porosity and permeability fields (see Christie & Blunt, 2001). Here we used a grid size of $30 \times 60 \times 3$ m^3 with uniform grid-block volume of $12 \times 6 \times 1.2$ m^3. The injection and production wells were placed at opposite corners of the reservoir. A single component, n-decane, was used to model the oil and the EOS used was the SRK equation. Steam injection was modeled using an injection stream of 1 mol% CO_2 and 99 mol% steam while STRIP, which contains more CO_2 from combustion, injected 10 mol% CO_2 and 90 mol% water. The heat and water input were the same for steam injection and STRIP so an equitable comparison could be drawn.

Main Simulation Results

The performance of steam injection and STRIP are compared using the metrics of sweep efficiency and oil recovery.

Sweep Efficiency

The sweep efficiency can be deduced from oil saturation at the end of the operating period.

Fig. 3 shows the oil saturation in the reservoir for steam injection and STRIP after 2000 days of operation.

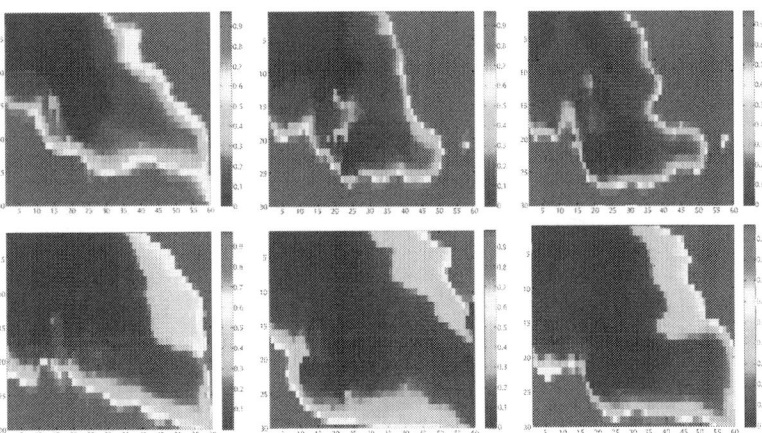

Figure. 3: Oil Saturation for Steam Injection (left) and STRIP (right) for Example 5.

Note that the blue regions are much larger for STRIP than for steam injection, indicating that STRIP removes more oil. One can also make more quantitative measures of sweep efficiency using the following expression

$$\eta = \frac{V_{oil}^{\Delta}}{V}$$

(29)

Where η denotes the sweep ratio, V_{Oil}^{Δ} is the porous volume for which the oil composition has changed by 1% or more and V is the total porous volume available to the oil. Fig. 4 shows quantitative results for sweep ratio for steam injection and STRIP as a function of time. The sweep ratios for steam injection and STRIP after 2000 days of operation are 60% and 83%, respectively. Clearly the sweep ratio of STRIP is superior to steam injection.

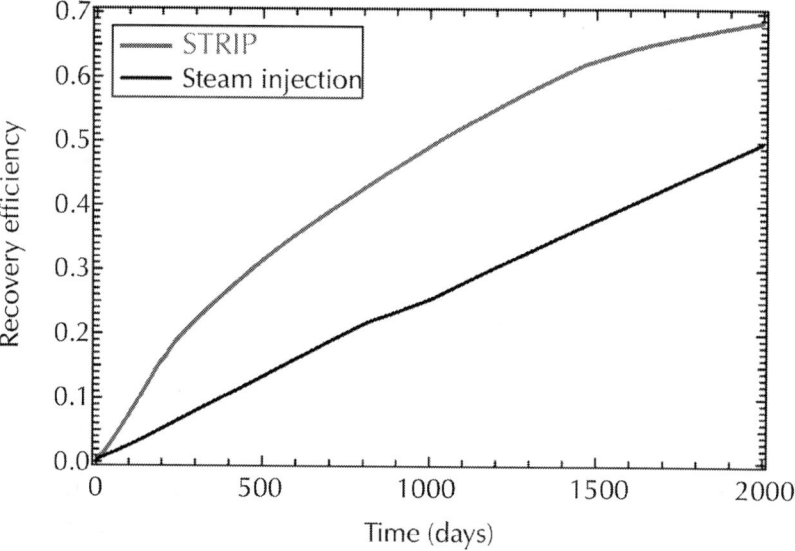

Figure. 4: Sweep ratio for steam injection and STRIP for Example 5.

Oil Production

Fig. 5 shows the total cumulative oil recovered during 2000 days of operation of steam injection and STRIP.

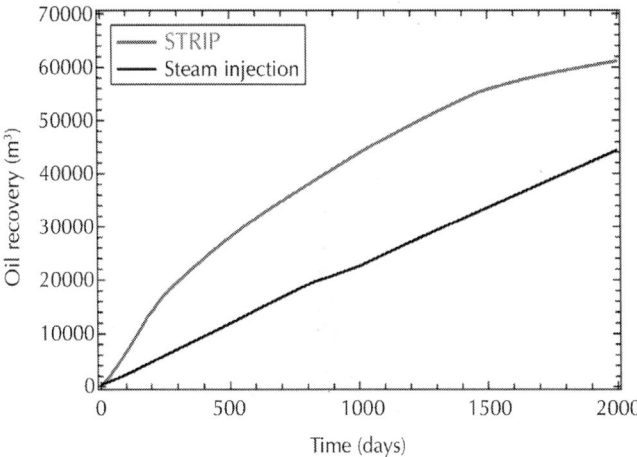

Figure. 5: Oil Recovery for Steam Injection and STRIP for Example 5.

To compare oil recovery, the Original Oil in Place (OOIP) at surface conditions, which is a common assumption in the petroleum industry, must be computed. OOIP is calculated using formula

$$\text{OOIP} = \sum_{\text{all blocks}} \frac{V\phi S_{oil}}{B_{oil}}$$

(30)

where V is the block volume, $Soil$ is oil saturation, and $Boil$ is the surface-to-reservoir formation volume factor. Here again, STRIP outperforms steam injection – recovering 16,764 m³ (105,442 barrels) more oil and leaving less OIP after 2000 days. Table 8 summarizes the performance of steam injection and STRIP for this first example.

Table 8: Summary of steam injection and STRIP performance for Example 5

	Steam injection	STRIP	Improvement with STRIP
Operation (days)*	2000	2000	
OOIP (m³)	89,051	89,051	

Oil produced (m³)	44,380	61,144	16,764
% oil recovered	50%	69%	38%
Sweep efficiency	60%	83%	27.4%

Example 6: Comparisons between Conventional EOS and CSAT

This second example compares a conventional reservoir simulation approach, which uses an EOS, to one that uses CSAT. For this example, pore volumes and permeability fields were taken from the upper layer of the original SPE10 model (Christie & Blunt, 2001). The simulations were performed using an initial reservoir composition of 1 mol% CO_2, 49 mol% n-decane, 20 mol% n-hexadecane, and 30 mol% water and the initial reservoir pressure and temperature were 31 bar and 300 K, respectively. One injection and one production well were placed at opposite corners of the reservoir. The injection well operates under constant pressure and temperature conditions of 60 bar and 500 K. The STRIP injection fluid consisted of 15 mol% CO_2, and 85 mol% water. The production well was set to a constant pressure of 3.45 bar. Input data for this example are shown in Table 9.

Table 9: Input data for STRIP simulation of Example 6

Quantity	Value
Reservoir dimensions	$365 \times 670 \times 0.6096$ m³
Initial reservoir T & p	300 K, 31 bar
Initial reservoir composition	1 mol% CO_2, 49 mol% C_{10}, 20 mol% C_{16}, 30 mol% water
Average porosity	0.1945
Permeability (SPE10, upper layer)[a]	8 orders of permeability variations
Injection T & p	500 K, 60 bar

Injection composition	15% CO_2, 85% water
Production well p	3.45 bar
Time horizon	7000 days

Main Simulation Results

The details of the performance of STRIP are discussed along with the features of the simulator.

Sweep Efficiency

Oil and gas saturations provide enough information to quantify sweep efficiency. Fig. 6 shows the oil and gas saturation in the reservoir for STRIP after 7000 days of operation, where the x and y axes denote grid blocks and the color bar shows saturations.

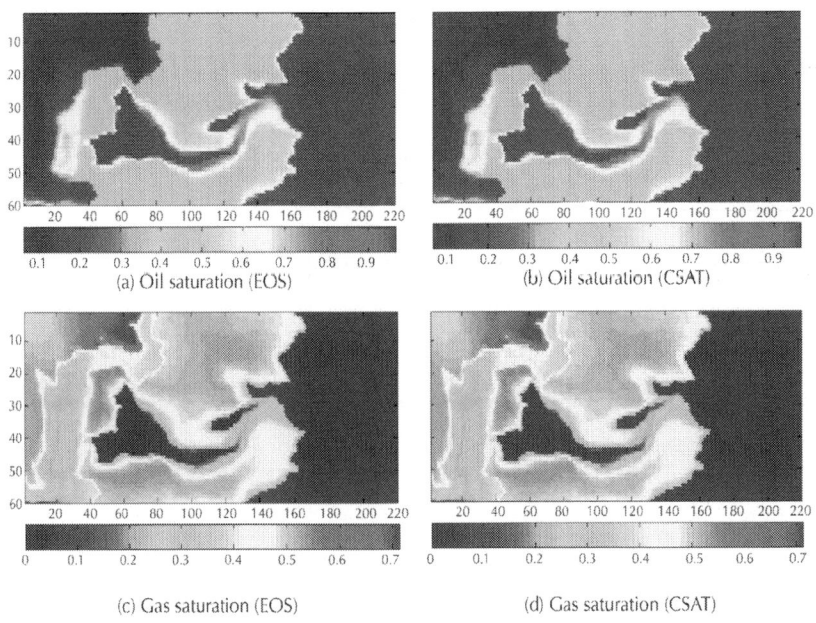

Figure. 6: Oil and gas saturation distributions for EOS and CSAT simulations after 7000 days.

As expected, both CSAT and GFLASH provide identical results for gas and oil saturation. This is because CSAT only skips phase identification and rigorous flash computations when compositions are far from phase boundaries.

Simulation Statistics

Table 10 summarizes the simulation statistics for this example and shows that rigorous flash solutions take the bulk of the computer time for the conventional EOS approach. CSAT, on the other hand, significantly decreases the number of EOS solves and, therefore, reduces total flash solution time by almost two orders of magnitude.

Table 10: Statistics for STRIP reservoir simulation of Example 6

	AD-GPRS/GFLASH	AD-GPRS/GFLASH/CSAT
Time horizon (days)	7000	7000
Average time step (days)	4.9	4.9
Model formulation	Natural Variables	Natural Variables
No. of grid blocks	13,200	13,200
No. of equations/ grid block	2 + P-1 + (C-1)P	2 + P-1 + (C-1)P
Equation solving methodology	Fully Implicit Method (FIM)	Fully Implicit Method (FIM)
Total No. of Newton iterations	4140	4140
Total No. of EOS solves	44,028,736	499,693
Total flash solution time (CPU sec)	270,830	3869
Total simulation time (CPU sec)	345,659	79,710

CONCLUSIONS

A new methodology for reservoir simulation was presented. This new modeling and simulation framework consists of AD-GPRS, the Automatic

Differentiation – General Purpose Research Simulator, a general multi-phase equilibrium flash suite, GFLASH, and a Compositional Space Adaptive Tabulation (CSAT) approach. The fundamental PDE model equations and methods of solution for the resulting nonlinear algebraic equations at the reservoir scale were provided. Modeling, equation-solving, and numerical details for four separate chemical engineering problems at the flash level of the computations were also presented to raise awareness in the PSE community with regard to reservoir simulation. Coupling of the flash and reservoir equations was described. CSAT and the interface between AD-GPRS, which is written in C++, and GFLASH, which is a FORTRAN program suite, were also described. Two numerical reservoir simulation examples were presented to highlight the accuracy, reliability, and computational efficiency of AD-GPRS/GFLASH, including two three-phase reservoir simulation examples with and without the use of CSAT for a highly heterogeneous reservoir formation and for three- and four-component system. Comparisons of steam injection and STRIP in Example 5 clearly demonstrate the superiority of the Solvent Thermal Resource Innovation Process in terms of sweep and oil recovery. Example 6 demonstrates that the AD-GPRS/GFLASH/CSAT framework reduces the simulation time by two orders of magnitude without losses in accuracy or reliability.

CODA

In addition to the chemical engineering sub-problems described in this work, there are also others, including sub-problems that require

- The characterization of oils with many components.
- Development of better methods for determining viscosity and relative permeability in harsh conditions.
- Understanding chemical EOR methods and the associated phase equilibrium in the presence of surfactants and other chemical additives.
- The determination of asphaltene precipitation.
- Reaction kinetics models for gas hydrate formation and CO_2 sequestration.

- Improved numerical methods for flash and for solving 'stiff' DAE systems.

In our opinion, the PSE community is ideally positioned to make contributions in these and other areas as they relate to reservoir simulation.

REFERENCES

1. Acs, G. (1985). General purpose compositional model. Journal of Petroleum Science and Engineering, 25, 543.
2. Ahlers, J., & Gmehling, J. (2001). Development of a universal group contribution equation of state. 1. Prediction of liquid densities for pure compounds with a volume translated Peng-Robinson equation of state. Fluid Phase Equilibria, 191, 177.
3. Ahlers, J., & Gmehling, J. (2002). Development of a universal group contribution equation of state. 2. Prediction of vapor-liquid equilibria for asymmetric systems. Industrial & Engineering Chemistry Research, 41, 3489.
4. Aziz, K., Ramesh, A. B., & Woo, P. T. (1987). Fourth SPE comparative solution project: comparison of steam injection simulators. JPT, 39, 1576.
5. Cao, H.(2002). Development oftechniques for general purpose simulators (Ph.D.thesis). Stanford University.
6. Christie, M., & Blunt, M. (2001). Tenth SPE comparative solution project: A comparison of upscaling techniques. SPE Reservoir Evaluation and Engineering, 4, 308.
7. Coats, K. H. (1980). An equation of state compositional model. Journal of Petroleum Science and Engineering, 1, 363.
8. Coats, K. H., Nielsen, R. L., Terhune, M. H., & Weber, A. G. (1967). Simulation of three-dimensional, two-phase flow in oil and gas reservoirs. Transactions of the American Institute of Mining and Metallurgical Engineers, 237, 377.
9. Douglas, J., Peaceman, D. W., & Rachford, H. H.(1959). A method of calculating multidimensional immiscible displacement. Transactions of the American Institute of Mining and Metallurgical Engineers, 216, 297.

10. Holderbaum, Th., & Gmehling, J. (1991). PSRK: A group-contribution equation of state based on UNIFAC. Fluid Phase Equilibria, 70, 251.

11. Iranshahr, A., Voskov, D. V., & Tchelepi, H. A. (2010). Generalized negative-flash method for multiphase multicomponent systems. Fluid Phase Equilibria, 2, 299.

12. Kiepe, J., Horstmann, S., Fischer, K., & Gmehling, J. (2004). Application of the PSRK model for systems containing strong electrolytes. Industrial Engineering Chemistry Research, 43, 6607.

13. Lucia, A., Bonk, B. M., Waterman, R. R., & Roy, A. (2012). A multi-scale framework for multi-phase equilibrium flash. Computers & Chemical Engineering, 36, 79.

14. Lucia, A., & Feng, Y. (2003). Multivariable terrain method. AIChE Journal, 49, 2553.

15. Marcondes, F., Maliska, C. R., & Zambaldi, M. C. (2009). A comparative study of implicit and explicit methods using unstructured Voronoi meshes in petroleum reservoir simulation. Journal of the Brazilian Society of Mechanical Sciences and Engineering, 31, 353.

16. Michelsen, M. L. (1982a). The isothermal flash problem. Part I. Stability. Fluid Phase Equilibria, 1, 9.

17. Michelsen, M. L. (1982b). The isothermal flash problem. Part II. Phase-split calculation. Fluid Phase Equilibria, 1, 9.

18. Odeh, A. S. (1981). Comparison of solutions to a three-dimensional black oil reservoir simulation problem. JPT, 33, 13.

19. Ortega, J. M., & Rheinboldt, W. C. (1970). Iterative solution of nonlinear equations in several variables. Philadelphia, PA: SIAM.

20. Peaceman, D. W. (2000). Fundamentals of numerical reservoir simulation. Amsterdam, The Netherlands: Elsevier Scientific Publishing Co.

21. Peery, J. H., & Herron, E. H. (1969). Three-phase reservoir simulation. Transactions of the American Institute of Mining and Metallurgical Engineers, 246, 211.

22. Peneloux, A., Rauzy, E., & Freze, R. (1982). A consistent correction for Redlich-KwongSoave volumes. Fluid Phase Equilibria, 8, 7.

23. Peng, D. Y., & Robinson, D. B. (1976). A new two-constant equation of state. Industrial Engineering Chemistry Fundamentals, 15, 59.

24.   Price, H. S., & Coats, K. H. (1974). Direct methods in reservoir simulation. Transactions of the American Institute of Mining and Metallurgical Engineers, 257, 295.

25.   Reid, R., Prausnitz, J. M., & Poling, B. E. (1987). The properties of gases and liquids (4th ed.). New York, NY: McGraw-Hill Co.

26.   Sandler, S. I. (1999). Chemical and engineering thermodynamics (3rd ed.). New York, NY: John Wiley & Sons, Inc.

27.   Settari, A., & Aziz, K. (1974). A computer model for two-phase coning simulation. Journal of Petroleum Science and Engineering, 14, 221.

28.   Sheffield, M.(1969). Three-phase fluid flow including gravitational, viscous and capillary forces. Transactions of the American Institute of Mining and Metallurgical Engineers, 257, 232.

29.   Snyder, L. J. (1969). Two-phase reservoir flow calculations. Journal of Petroleum Science

30.   and Engineering, 9, 170.

31.   Soave, G. (1972). Equilibrium constants from a modified Redlich-Kwong equation of state. Chemical Engineering Science, 27, 1197.

32.   Todd, M. R., O'Dell, P. M., & Hirasaki, G. J. (1972). Methods for increased accuracy in numerical reservoir simulation. Transactions of the American Institute of Mining and Metallurgical Engineers, 253, 515.

33.   Trimble, R. H., & McDonald, A. E. (1976). A strongly coupled, implicit well coning model. In Soc. Pet. Eng. 4th symposium on numerical simulation of reservoir performance Los Angeles, CA, SPE Paper No. 5738.

34.   Voskov, D., & Tchelepi, H. (2012). Comparison of nonlinear formulations for twophase multi-component EOS-based simulation. Journal of Petroleum Science and Engineering, 82-83, 101.

35.   Voskov, D., & Tchelepi, H. (2009a). Compositional space parameterization: Theory and application for immiscible displacements. Journal of Petroleum Science and Engineering, 14, 431.

36.   Voskov, D., & Tchelepi, H. (2009b). Compositional space parameterization: Multicontact miscible displacement and

extension to multiple phases. Journal of Petroleum Science and Engineering, 14, 441.

37. Voskov, D., & Zhou, Y.(2012). Technical description of AD-GPRS. Stanford, CA: Stanford

38. University.

39. Whitson, C., & Michelsen, M. L. (1989). The negative flash. Fluid Phase Equilibria, 53, 51.

40. Zaydullin, R.,Voskov, D.V., & Tchelepi, H.A.(2013). Formulation and solution of compositional displacements in tie-simplex space. In 2013 SPE reservoir simulation symposium Society of Petroleum Engineers, The Woodlands, TX.

41. Zhou, Y., Tchelepi, H. A., & Mallison, B. T. (2011). Automatic differentiation framework for compositional simulation on unstructured grids with multi-point discretization schemes. In SPE reservoir simulation symposium SPE141592, February 2011.

Chapter 4

Methanol treatment in Gas Condensate Reservoirs: A Modeling and Experimental Study

A. Asgari[a], M. Dianatirad[a], M. Ranjbaran[b], A.R. Sadeghi[a], And M.R. Rahimpour[a, c]

[a]Department of Chemical Engineering, School of Chemical and Petroleum Engineering, [b]Shiraz University, Shiraz 71345, Iran

Department of Chemical and Petroleum Engineering, Sharif University of Technology, Tehran, Iran

[c]Gas Center of Excellence, Shiraz University, Shiraz 71345, Iran

ABSTRACT

Well productivity in gas condensate reservoirs is reduced by condensate blockage when the bottom-hole pressure drops below dew point pressure. The present experimental study on limestone cores shows that the relative permeability of gas decreases due to liquid blockage; furthermore, methanol has proven effective in the removal of condensate and restoration of gas relative permeability. In this research, the decrease in gas relative permeability caused by condensate banking and the effect of methanol treatment on condensate-blocked rocks was simulated using the cubic-plus-association (CPA) equation of state. The CPA equation of state was applied to the modeling of two-phase flows through cores for methanol–hydrocarbon mixtures due to charge transfer and hydrogen bonding, both of which may strongly affect the thermodynamic properties of such mixtures. Differential equations were solved by means of the orthogonal collocation method, a method particularly attractive for solving nonlinear problems. The modeling results confirm the experimental results, and both methods indicate that significant productivity loss can occur in retrograde gas condensate reservoirs when the flowing bottom-hole pressure falls below dew point pressure. Moreover, the results show that methanol treatment can improve gas relative permeability by a factor of about 1.3–1.6. These results may help reservoir engineers and specialists to restore the lost productivity of gas condensate.

INTRODUCTION

At present, natural gas reservoirs are one of the world's main sources of energy, accounting for approximately a quarter of worldwide energy demand. It is also worth noting that global demand for natural gas has been growing rapidly in recent years (BP, 2013). According to recent figures provided by the International Energy Agency (IEA), world gas consumption is expected to rise by 1.5% per annum by 2030. Many of the largest natural gas reservoirs have retrograde properties, which result in liquid accumulation near the wellbore due to pressure drop occurring during the production of gas. This formation of liquid around the wellbore reduces gas relative permeability and thus well productivity. The phase behavior of a gas condensate reservoir is strongly dependent on the $P–T$ envelope and thermodynamic conditions of the hydrocarbon mixture. Gas condensate reservoirs generally produce gas in the range of 30–300 STB/MMSCF (standard barrels of liquid per million standard cubic feet of gas). In addition, the ranges of pressure (P) and temperature (T) for this type of gas reservoir are usually between 3000 and 8500 psi and 150–400 °F, respectively (Zendehboudi et al., 2012).

Fevang and Whitson (1995) have characterized retrograde gas reservoirs to exhibit three different regions. Region 1 is the part around the wellbore where condensate can flow, while region 2 is the part of the reservoir where condensate begins to form but cannot flow. Region 3, on the other hand, is the mid-to-outer boundary of the reservoir where only single-phase gas exists.

Several methods have been proposed to improve gas relative permeability in the event of condensate aggregation around the wellbore. Gas injection (Abel et al., 1970, Kossack and Opdal, 1986, Sänger and Hagoort, 1998 and Hoier et al., 2004) and water-altering gas (Cullick et al., 1993, Henderson et al., 1991,Jones et al., 1993 and Fishlock and Probert, 1996) are two methods used to maintain reservoir pressure above dew point pressure. However, these two methods are not economical due to the large initial investment required and higher operational costs involved (Ahmed et al., 2000). Hydraulic fracturing and horizontal wells have also been used to enhance gas productivity (Settari et al., 1996, Al-Hashim and Hashmi, 2000, Kumar, 2000, Lolon et al., 2003 and Mohan, 2005). By inducing a hydraulic fracture, the bottom-hole pressure and area available for gas and condensate flow can be increased. Nonetheless, the success of hydraulic fracture stimulation depends on many parameters, such as reservoir permeability, fluid composition, proppant volume, and the degree to which the fracture cleans up after the treatment. Many researchers have also proposed chemical-based treatments. It has been shown that altering wettability from oil-wet to intermediate gas-wet leads to reduced oil saturation (Jadhunandan and Morrow, 1991, Owolabi and Watson, 1993 and Chen et al., 2004). Li and Firoozabadi were the first who proposed the enhancement the gas deliverability via altering wettability using a phenomenological model and laboratory experiments in gas condensate reservoirs. They also succeeded in altering the wettability of Berea sandstone and Kansas chalk from water-wet to intermediate gas-wet using various chemicals at room temperature (Li and Firoozabadi, 2000a, Li and Firoozabadi, 2000b and Bang et al., 2010). Gilani et al. (2011) performed similar experiments on sandstone and limestone cores. However, chemical treatment, particularly non-ionic surfactant on limestone, does not cause significant improvement in gas relative permeability. There is a clear need for an effective treatment solution for carbonate rocks as many of the world›s hydrocarbon reservoirs, including those in

Iran, are based on carbonate rock formation. Nevertheless, methanol treatment has proved effective for this type of reservoir rock. The use of an inexpensive solvent such as methanol to improve the productivity of gas condensate reservoirs presents an attractive approach. Walker (2000) and Du et al. (2000) investigated the applicability of methanol treatment to improve the productivity of gas condensate fields. Their research revealed that the removal of the condensate bank is temporary, and that the formation of a condensate bank does not occur immediately. They proposed that the residual methanol in the pores delayed the reformation of the condensate bank. Al-Anazi et al., 2002 and Al-Anazi et al., 2003 studied the effect of methanol on limestone and sandstone cores. Their study showed that methanol can displace water and condensate and improve the relative permeability of gas. Al-Anazi et al. (2005) reported a successful case of methanol injection in Alabama (Hatter›s Pond Field). In this study, methanol was injected at a rate of 8 bbl/min, resulting in increases in both gas and condensate production rates by a factor of 2 over the first 4 months and by 50% thereafter. Alzate et al. (2006) investigated the effect of alcohol-based and inhibited-diesel on the gas effective permeability on both Mirador formation in Cupiagua Main Field, Colombia and Berea sandstone. They showed that alcohol labeled 21-NE-06 and inhibited-diesel increase the gas effective permeability. The properties of gas and condensate flow when pressure falls below dew point are significantly different from those of conventional gas–oil systems. Muskat (1949) and Fetkovich (1973)modeled gas condensate reservoirs and presented a simple method for estimating the radius of condensate blockage as a function of time, gas rate, and reservoir rock and fluid properties. Narayanaswamy, 1998 and Narayanaswamy, 1999 proposed an analytical approach to calculate the non-Darcy flow coefficient for heterogeneous reservoirs. Kniazeff and Naville (1965) and Eilerts et al. (1965)were the first to numerically model radial gas-condensate well deliverability. These studies represent radial saturation and pressure profiles as functions of time and other operational variables, confirming that condensate blockage indeed reduces well deliverability. Fevang and Whitson (1995) presented an accurate method for modeling the deliverability of gas condensate wells. In this study, well deliverability was calculated using a modified version of the Evinger–Muskat pseudo-pressure model, with the gas condensate reservoir being divided to three flow regions. Mott (2003) proposed a novel technique to estimate gas condensate well production performance using the pseudo-pressure model. Bonyadi et al. (2012) also presented a new method for the prediction of condensate well productivity. The model was tested and compared with the results of the fine-grid simulation of two cases, namely rich and lean gas condensate fluids. In the present study, the effect of methanol on the phase behavior of reservoir fluids at reservoir conditions was simulated; moreover, two-phase flow equations in cylindrical coordination across the radial and axial orientations were solved. What distinguishes this investigation from previous works is the use of an attractive yet seldom applied method called orthogonal collocation for the solution of reservoir nonlinear equations and the application of the cubic-plus-association (CPA) equation of state, which is the most effective equation of state for the prediction of the phase behavior of alcohol–hydrocarbon mixtures. The cubic-plus-association (CPA) model is an equation of state that combines the cubic SRK equation of state, an association (chemical) term coined by Kontogeorgis et al. (1996).

MODELING

Governing Equations

Continuity equations are applicable to any flow system in reservoirs. Darcy's equation also needs to be applied in order to relate the fluid flow rate to reservoir pressure. For two-phase flows in porous media, the general partial differential equations are as follows (Ahmed, 2001):

Darcy's law:

$$u = -\frac{k}{\mu}\nabla P$$

(1)

Total mole balance:

$$\frac{\partial}{\partial t}\left[\varphi\left(\rho_g S_g + \rho_l S_l\right)\right] - \nabla \cdot \left[\left(\rho_g \frac{k_g}{\mu_g} + \rho_l \frac{k_l}{\mu_l}\right)\nabla P\right] = 0 \qquad (2)$$

Component mole balance:

$$\frac{\partial}{\partial t}\left[\varphi\left(\rho_g S_g y_i + \rho_l S_l x_i\right)\right] - \nabla \cdot \left[\left(\rho_g \frac{k_g}{\mu_g} y_i + \rho_l \frac{k_l}{\mu_l} x_i\right)\nabla P\right] = 0 \qquad (3)$$

Equilibrium relations are solved as a set of K-value problems in the form

$$y_i = K_i x_i \qquad (4)$$

For the last component:

$$\left(1 - \sum_{i=1}^{nc-1} y_i\right) = K_{nc}\left(1 - \sum_{i=1}^{nc-1} x_i\right) \qquad (5)$$

Where yi and xi are the mole fractions of component i in the gas and condensate phases, respectively. Ki is calculated using the gas and condensate phase fugacity.

Equation of State

Species-forming bonds often exhibit unusual thermodynamic behavior. There are strong attractive interactions between molecules of the same species (self-association) and between molecules of different species (cross-association) (Yakoumis et al., 1997, Yakoumis et al., 1998 and Solms et al., 2004). CPA is an equation of state that combines the cubic SRK equation of state (physical) and association (chemical) terms. The chemical term in the CPA equation of state, similar to the SAFT equation, is obtained from the Wertheim first-order perturbation theory. In this work, CPA was applied to the modeling of vapor–liquid equilibrium for alcohol–hydrocarbons. For the determination of liquid density, we added the Peneloux volume shift to the equations (Danesh, 1998). In terms of the compressibility factor Z, the CPA equation is presented as:

$$Z = Z^{SRK} + Z^{assoc} \tag{6}$$

The *P*-explicit form of the CPA equation of state is as follows:

$$P = \frac{RT}{v-b} - \frac{a}{v(v+b)} - \frac{1}{2}\frac{RT}{v}\left(1 + \frac{1}{v}\frac{\partial \ln g}{\partial(1/v)}\right)\sum_{i=1}^{nc} x_i \sum_{A_i}\left(1 - X_{A_i}\right) \tag{7}$$

$$\ln \hat{\varphi}_i = \frac{b_i}{b}(Z-1) - \ln\ (Z-\beta) - \tilde{q}_i I$$

$$-\frac{1}{2}\frac{\partial \ln g}{\partial n_i}\sum_{j=1}^{nc} x_j \sum_{A_j}\left(1 - X_{A_j}\right) \tag{8}$$

The method of solution is the same as that mentioned by Smith et al. (2005). The only difference is the presence of an association term, which is placed in the outer cycle of trial and error.

$$g = \frac{1}{1-1.9\eta} \tag{9a}$$

$$\eta = \frac{b}{4v} \tag{9b}$$

Where g is the radial distribution function, which is representative of the molecules' microscopic structures that are further simplified by Kontogeorgis et al. (1999), and   is the reduced fluid density.

A mixture of alcohol and hydrocarbon contains not only monomer species, but also associated clusters. Thus X_{Ai} is defined as the mole fraction of molecules i that are not bonded at site A. Haung and Radosz (1990) have identified eight different association schemes which can be applied to different molecules depending on the number and type of association sites. For alcohols, each OH group has three association sites labeled A and B on oxygen and C on hydrogen. Due to similar O–O or H–H (AA, AB, BB, and CC) interactions, the association strength (Δ) is assumed to be equal to zero. Attraction can only occur between AC and BC. With respect to self-association, the association schemes of methanol and water are 2B and 4C, respectively. Table 1 shows the equations for pure component and self-association as presented by Huang and Radosz.

Table 1: Association schemes presented by Haung and Radosz (1990)

Type	Δ	X	XA
1	$\Delta AA \neq 0$	–	$\dfrac{-1+\sqrt{1+4\rho\Delta}}{2\rho\Delta}$
2A	$\Delta AA = \Delta AB = \Delta BB \neq 0$	XA = XB	$\dfrac{-1+\sqrt{1+8\rho\Delta}}{4\rho\Delta}$
2B	$\Delta^{AA} = \Delta^{AB} = 0$ $\Delta^{BB} \neq 0$	XA = XB	$\dfrac{-1+\sqrt{1+4\rho\Delta}}{2\rho\Delta}$
3B	$\Delta^{AA} = \Delta^{AB} = \Delta^{BB} = \Delta^{CC} = 0$ $\Delta^{AB} = \Delta^{BC} \neq 0$	XA = XB	$\dfrac{-(1-\rho\Delta)+\sqrt{(1-\rho\Delta)^2+4\rho\Delta}}{4\rho\Delta}$
		XC = 2XA−1	
4C	$\Delta^{AA} = \Delta^{AB} = \Delta^{BB} = \Delta^{CC} =$ $\Delta^{CD} = \Delta^{DD} = 0$ $\Delta^{AC} = \Delta^{AD} = \Delta^{BC} = \Delta^{BD} \neq 0$	XA = XB = XC = XD	$\dfrac{-1+\sqrt{1+8\rho\Delta}}{4\rho\Delta}$

Fig. 1 shows the type of bounding in real associating fluids.

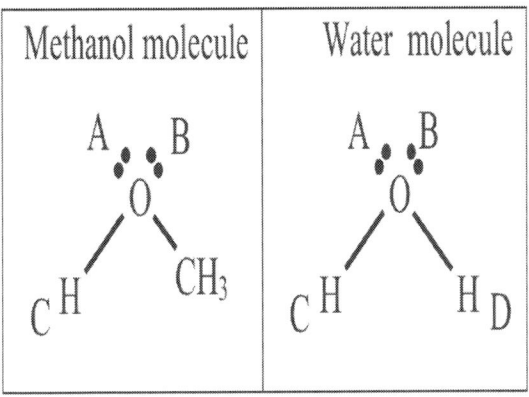

Figure 1: Types of bonding in real associating fluids.

ΔAiBj denotes the association between site A on molecule *i* and site B on molecule *j*.

$$X_{A_i} = \cfrac{1}{1 + \rho \displaystyle\sum_{i=1}^{nc} x_j \sum_{B_j} X_{B_j} \Delta^{A_i B_j}} \tag{10}$$

$$\Delta^{A_i B_j} = g \left[\exp \left(\frac{\varepsilon^{A_i B_j}}{RT} \right) - 1 \right] b_{ij} \beta^{A_i B_j} \tag{11a}$$

$$b_{ij} = \frac{b_i + b_j}{2} \tag{11b}$$

For cross-association, use is made of the Elliot combination rule (Elliott et al., 1990).

$$\Delta^{A_i B_j} = \sqrt{\Delta^{A_i B_i} \Delta^{A_j B_j}} \tag{12}$$

$$\varepsilon^{A_i B_j} = \frac{\varepsilon^{A_i B_i} + \varepsilon^{A_j B_j}}{2} \tag{13a}$$

$$\beta^{A_i B_j} = \sqrt{\beta^{A_i B_i} \beta^{A_j B_j}} \tag{13b}$$

The algorithm for the calculation of fugacity using the CPA equation of state appears in Fig. 2.

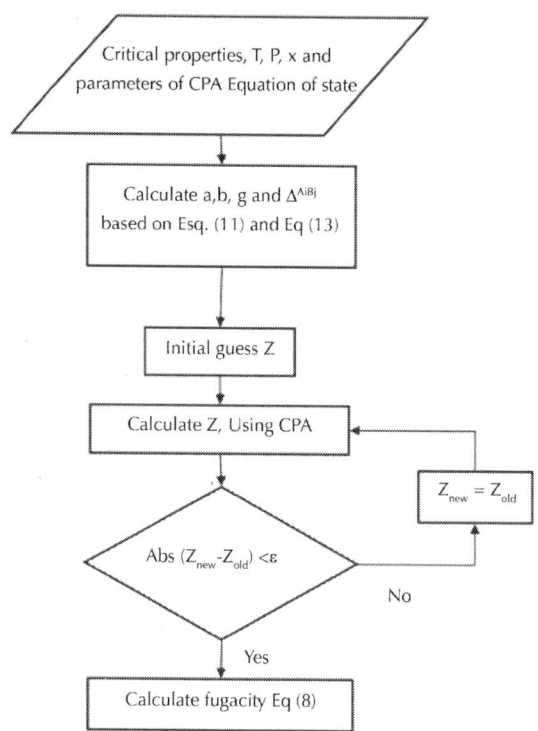

Figure 2: Algorithm for fugacity calculation for CPA model.

Orthogonal Collocation Method

The collocation method is one of the five widely used variations of the method of weighted residuals for engineering applications. It is distinguished by the choice of the test functions used in the minimization of the residuals (Rice and Do, 1995). The test function is the Dirac delta function at N interior points (called collocation points) within the domain of interest.

A basic procedure for the simulation of reservoirs using the orthogonal collocation method is shown in Fig. 3.

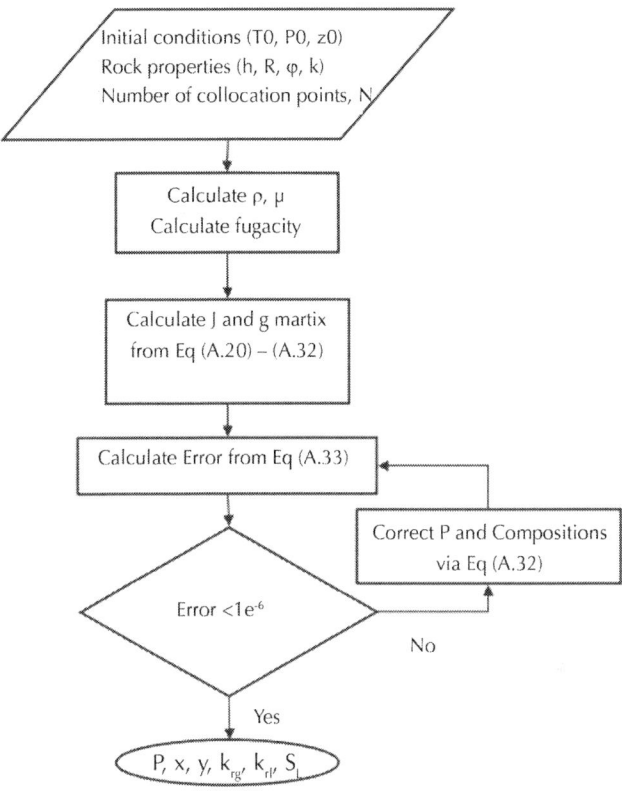

Figure 3: A basic procedure for simulation of the reservoirs using orthogonal collocation method.

MODEL VALIDATION

V–L phase equilibrium was applied for the methanol–hydrocarbon mixtures. The performance of the CPA and SRK equations of state were tested on the correlation of the saturated vapor pressure of pure methanol, and the results were compared with the reference data in Fig. 4 (Young, 1910). As can be seen in this figure, the CPA model demonstrates better correlation than the SRK model. At higher temperatures, the CPA and SRK models yield similar results, yet at lower temperatures significant error can be observed in the results obtained using the SRK model. The saturated liquid densities obtained from these models were also compared with the reference data in Fig. 5 (Young, 1910), ate the saturated temperature between 50 °C and 240 °C. We can see that the SRK model failed to give results, which fit well with the reference data in liquid phase density. Generally, it can be said that the CPA model yields results, which enjoy higher correlation with the experimental data for associating compounds than the SRK equation of state.

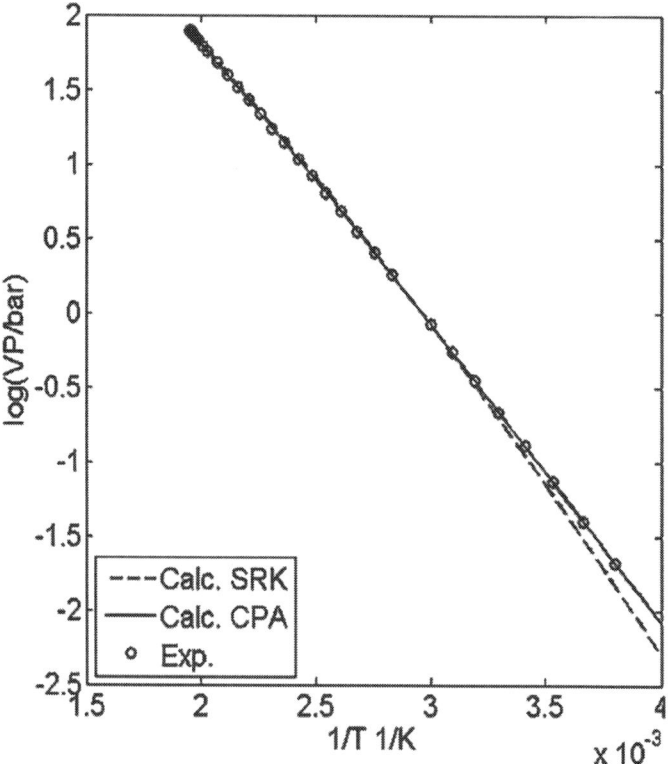

Figure 4: P–T diagrams for pure methanol using CPA and SRK EoS and experimental data (Young, 1910).

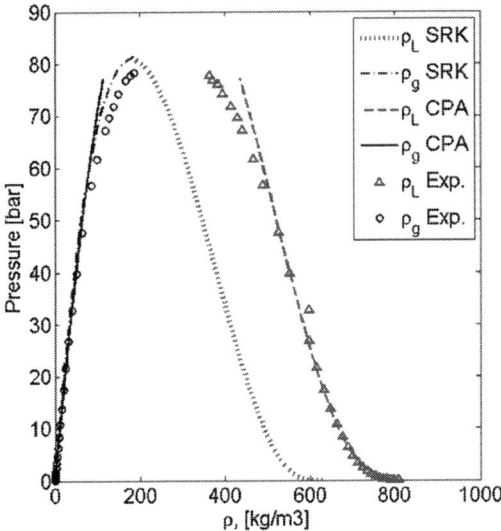

Figure 5: Density of pure methanol versus pressure using CPA and SRK EoS and experimental data (Young, 1910).

Fig. 6 shows that there was good agreement between the experimental data (Kontogeorgis et al., 1996) and the data obtained by means of modeling based on the CPA model in the case of the second virial coefficient.

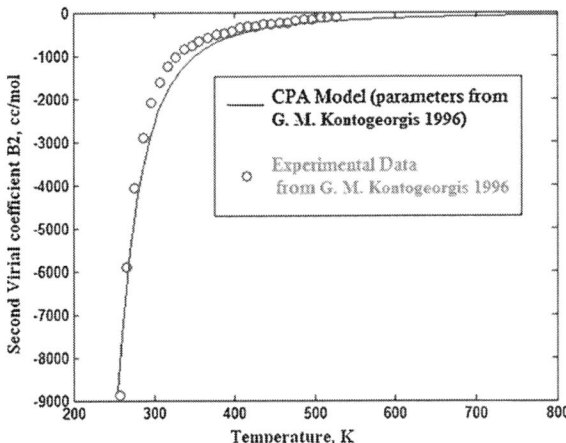

Figure 6: Second virial coefficient versus temperature using CPA EoS and experimental data given by (Kontogeorgis et al., 1996).

The effectiveness of the CPA model with respect to the p–x, y diagram for methanol–propane systems was also assessed. As illustrated in Fig. 7, a significant difference does not exist between the results of our calculations and those given in the reference paper by Kontogeorgis et al. (1996). Consequently, it can be said that the calculations carried out enjoy high correlation for systems consisting of methanol and hydrocarbon.

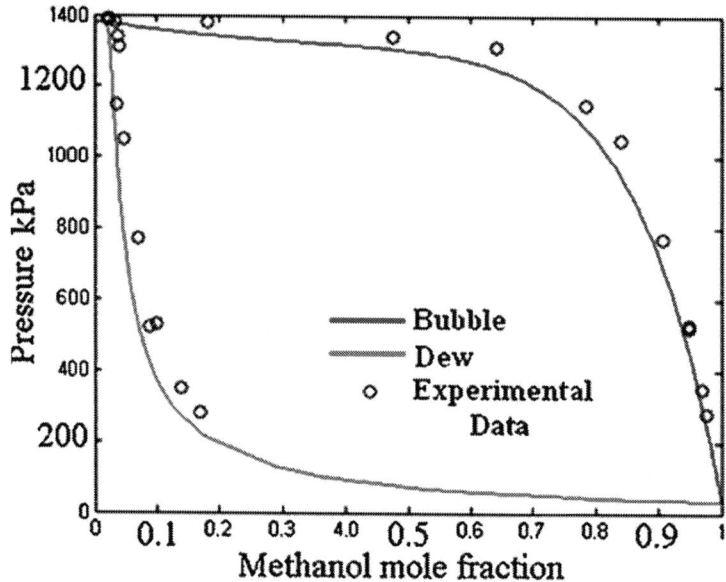

Figure 7: P–x, y diagrams for methanol/propane system from experimental data (Kontogeorgis et al., 1999) and calculated at 313.55 K using CPA equation of state.

EXPERIMENTAL STUDY

Core Flood Experimental Set Up

Core flood experiments were conducted on Asmari–Jahrom limestone cores. Fig. 8 shows a schematic of a gas condensate core flood apparatus. The core flood was designed for 10,000 psi and 350 °F. A

Vinci pump was used to inject fluid at a constant rate. In addition, a Vinci type core holder, with a diameter of 1.5 in. and variable length of up to 8 in. and could be opened and closed only at one end, was used in the experiments. The core holder material was stainless steel 316-L. An overburden pressure was applied using a hydraulic hand pump.

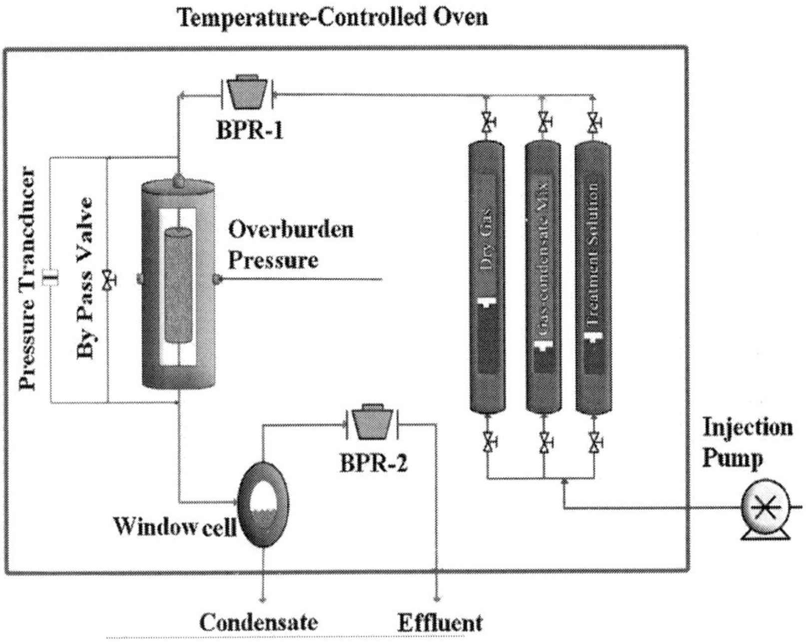

Figure 8: Schematic diagram of core flood setup.

Three piston accumulators were used for the storage of gas, condensate and solvents. Two backpressure regulators were used to control the pressure upstream and downstream of the core. A visual cell was installed in line to observe the fluid phases. The core holder, fluid accumulator, backpressure regulators and lines resided inside a temperature-controlled oven. Furthermore, a data acquisition system was used in this experiment in order to read data from the pressure transmitter and transfer them to a PC. This system was capable of digitally displaying 20 sets of data simultaneously. The recorded data were transferred to a computer using DP logger software.

Experimental Procedure

Core Preparation

At first, cores with 1.5-in. diameters and 8-in. lengths were cut from a source rock block. Two types of cores, namely Asmari–Jahrom and reservoir cores, were used. The cores were then dried in an oven at 160 °F. The porosity of the cores was measured using a Vinci helium porosimeter. Next, the cores were wrapped with aluminum foil and heat-shrink Teflon to prevent the diffusion of injected fluids through the Viton rubber sleeve. The wrapped core was subsequently placed into a core holder inside the oven at the experimental temperature. An overburden pressure was also applied using a hydraulic hand pump. Table 2lists the properties of the cores used in all the experiments.

Table 2: Core properties used in experimental tests.

Properties	1-Asmari–Jahrom limestone	2-Sarkhun limestone
Length (in.)	8	8
Diameter (in.)	1.5	1.5
Porosity (%)	24.6	15.2
Water saturation (%)	0–45	0
Absolute permeability (md)	6.85	0.9

Gas Mixture Preparation

Two types of synthetic gas condensate mixtures were prepared and used in these experiments. Table 3gives the composition of these two fluids. Each mixture was prepared in an accumulator at room temperature on mass basis. The accumulator was shaken to mix the fluids and then placed inside the oven at experimental conditions for 24 h until it reached equilibrium conditions. Fig. 9 shows the liquid dropout curve for Fluid 1 at 140 °F.

Table 3: Composition of gas mixtures.

Component	Fluid 1 (mole fraction)	Fluid 2 (mole fraction)
Methane	0.8	0.85
n-Butane	0.15	0
n-Heptane	0.035	0.15
n-Decane	0.015	0

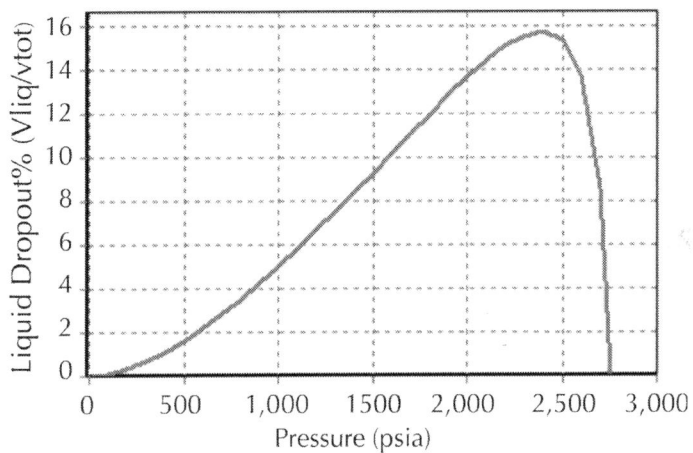

Figure 9: Simulated liquid dropout curve for Fluid 1 at 140 °F.

Stabilizing Water Saturations

In order to stabilize the initial water saturation, a Vinci de-saturator was used. The process involves the removal of water from the sample by suction wherein a porous ceramic wall serves as both a connecting link and a means of maintaining a difference in pressure between the liquid phase of the water in the soil and the water at lower pressure on the opposite side of the wall. The air pressure inside the chamber is raised, forcing excess water through the microscopic pores in the ceramic plate.

De-saturation times are subject to the permeability of samples. Typically, 2 or 3 days are allowed at each pressure point, as this duration is deemed adequate for most cases. After each pressure point, the sample is weighted and water saturation determined.

The multiple-sample desaturation cell facilitates the desaturation of a set of consolidated core samples using the porous plate method. Data obtained this way can be used to calculate capillary pressure versus saturation curves. A high-precision weighing balance can then be used for the purpose of core saturation measurement. The multiple-sample desaturation cell is manufactured from stainless steel and equipped with a large ceramic plate situated at its base and used as a semi-permeable porous plate. The setup consists of a capillary pressure cell with an easy-to-open lid, clamping bolts, O-ring seals, outflow tube assemblies, and a pressure control panel. Three ceramic plates capable of withstanding 3 bar, 5 bar and 15 bar of pressure are provided with the system. The control panel includes a digital pressure display, a pair of low- and high-range pressure regulators, and a set of control valves. A gas humidifier is also provided to insure minimum evaporation during the desaturation process.

Core Flood Procedure

The single-phase permeability of the cores was measured with methane at reservoir conditions in all the experiments. The outlet backpressure regulator was set at reservoir pressure; methane was then injected into the core at different flow rates using a Vinci pump until a steady state was reached. This experiment was repeated for each core and the absolute permeability was calculated from the steady state pressure drop across the core using Darcy's law.

To mimic the dynamic process of accumulation of condensate around the wellbore, the upstream backpressure regulator was set at a pressure above the dew point pressure and the downstream backpressure regulator was set at a pressure below the dew point pressure. Subsequently, the gas mixture was injected through the bypass line until the pressure in the lines stabilized. The bypass valve was then closed while the core holder inlet and outlet valves were opened. The process allowed the high-pressure gas mixture to flash at the inlet of the core and the condensate to accumulate dynamically inside the core. When a steady state of pressure drop was reached across the core, the gas and condensate relative permeabilities were calculated. Methanol was injected into the core at a rate of 2–5 cm³/min at experimental conditions to stimulate a reduction in gas production due to condensate buildup, water blockage, or the dual effect of water saturation and condensate banking. Methanol is miscible with both water and condensate at this condition. Finally, several pore volumes of methanol were injected and the longevity of the methanol was tested.

RESULTS AND DISCUSSIONS

Results of Modeling

Miscibility of Methanol and Hydrocarbon Mixtures

zTo investigate the phase behavior of methanol in a gas condensate reservoir, behavior studies were conducted using Fluid 1 with mixtures of methanol and Fluid 1. The pressure curve versus mole fraction of methanol has been plotted in Fig. 10. It is evident that at pressures above 1100 psi there is only one liquid phase, and the mole fraction of methanol in the gas phase is very low, meaning that methanol is a solvent which, at high pressures, is fully miscible with hydrocarbon liquids. These results validate the effectiveness of methanol injection, as condensate and methanol are miscible at reservoir conditions.

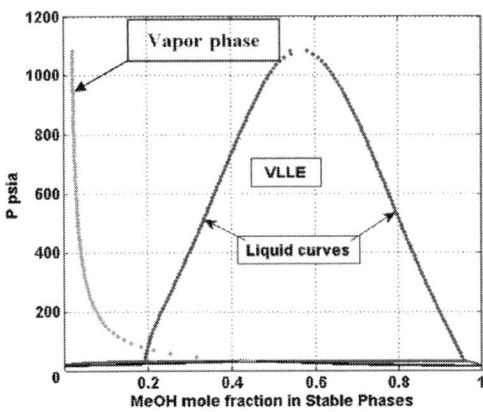

Figure 10: *P–x, y* diagrams for methanol/Fluid 1 systems.

Two-Phase Flow through Reservoir Cores

Pressure drop and gas relative permeability were calculated through the core for single-phase and two-phase flow following condensate liquid dropout for Fluid 2 using the CPA equation of state. Table 4 shows the core specifications and fluid properties used in the simulations. At 194 °F, the dew point of mixture (Fluid 2) is 2960 psi. The dew point pressure had to be determined accurately because the pressure needed to be set above the dew point pressure for the single-phase flow and below the dew point pressure for the two-phase flow in the calculations to simulate condensate blockage. Fig. 11 and Fig. 12 show the pressure drop across the core for the single-phase and two-phase flows before methanol injection, respectively.

Table 4: Core specifications and fluid properties used in simulations and experimental study for comparison

Core porosity (%)	16.7
Core length (mm)	205
Core diameter (mm)	38.1
Absolute permeability	0.92
Fluid initial pressure (psia)	3100
Fluid temperature(°F)	194
Maximum dropout (%)	12.1
Dew point pressure (psi)	2960
Gas flow rate (cm3/min)	5

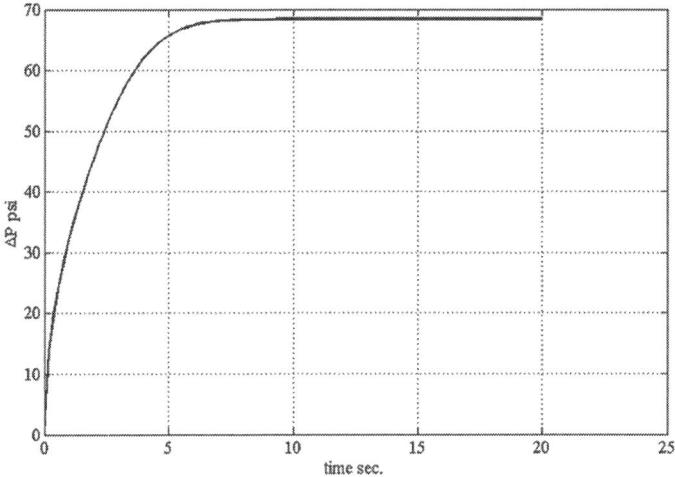

Figure 11: Overall core pressure drop for single-phase flow.

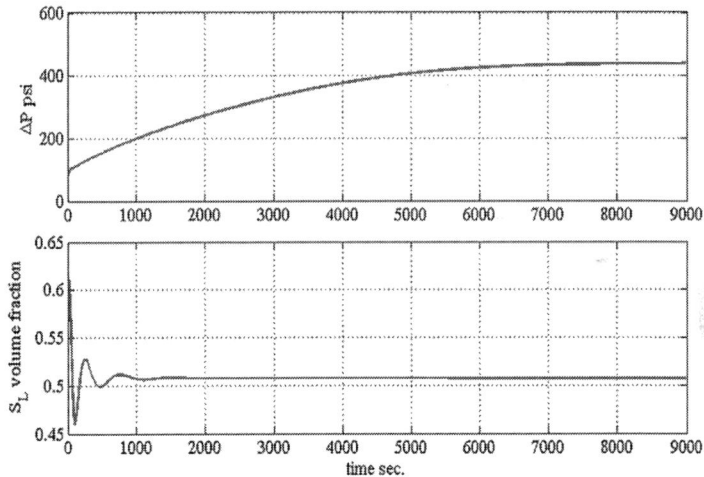

Figure 12: Condensate saturation and pressure drop across Sarkhun Limestone core during condensate accumulation at 1500 psig.

Fig. 12 illustrates that as pressure dropped below the dew point pressure, retrograde condensation occurred, leading to the formation of the condensate. The liquid phase accumulated in the core, causing a progressive decline in pressure across the core. Initially, the condensate phase started to fill up the pores while the gas phase was flowing. When the pressure drop reached a steady state, both gas and condensate reached steady state fractional values. Condensate saturation at steady state conditions increased to 0.51.

Fig. 13 depicts the gas and condensate relative permeabilities during the two-phase flow at 1500 psig. As can be seen from this figure, the gas relative permeability decreased to 0.166, whereas the condensate relative permeability increased to only 0.06. These results show that gas relative permeability and core total permeability are reduced due to condensate blockage. The buildup of a condensate bank in the core impedes the flow of gas and thus reduces its productivity or relative permeability.

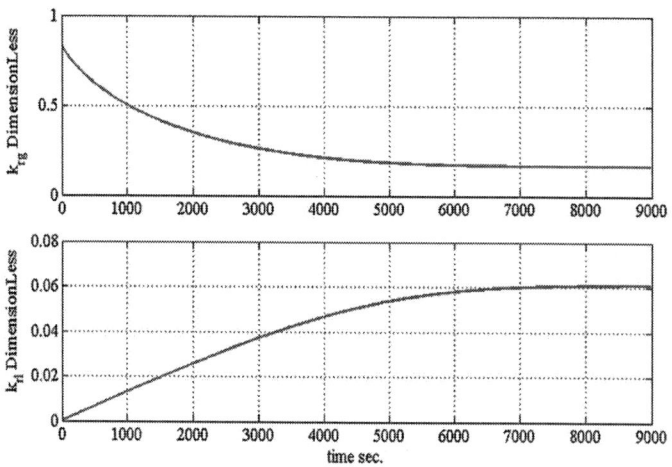

Figure 13: Gas and condensate relative permeability calculated for Sarkhun Limestone core during condensate accumulation at 1500 psig.

Methanol Injection

In the next part of the experiment, liquid methanol (100 cm^3) was injected into the core, and the pressure drop across the core was calculated as a function of time. The condensate accumulation step was then repeated at 1500 psig and the gas and condensate relative permeabilities were once again calculated. Fig. 14 shows the pressure drop across the core and condensate saturation before and after methanol injection. The use of methanol resulted in a fall in pressure drop and condensate saturation and thus higher gas relative permeability. Initially, pressure drop across the core increased after methanol injection due to the fact that some methanol existed in the pores. Afterwards, the methanol changed the phase behavior of the gas and condensate mixture in the core. Methanol is miscible with condensate; therefore, a likely reason for the increased gas relative permeability could be the miscible displacement of the condensate by the methanol. Fig. 15 shows the increase in gas relative permeability as a result of methanol injection. Subsequent to methanol injection, an increase in gas relative permeability from 0.166 to 0.231 was achieved. Compositional modeling shows that gas relative permeability increased to approximately 40%.

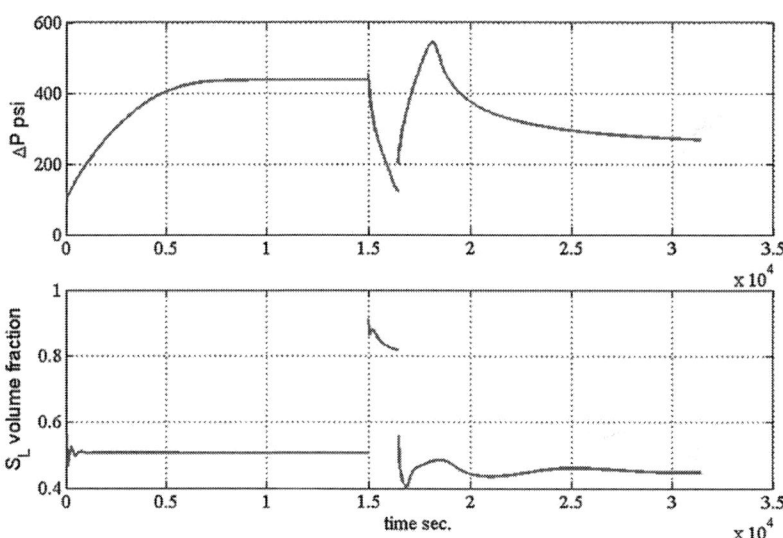

Figure 14: Condensate saturation and pressure drop across Sarkhun Limestone core before and after methanol injection at 1500 psig.

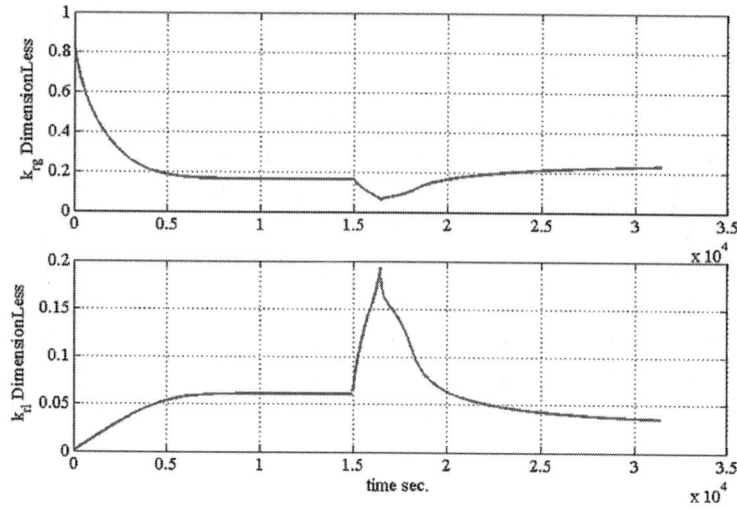

Figure 15: Gas and condensate relative permeability in Sarkhun Limestone core before and after methanol injection at 1500 psig.

Experimental Results

Pressure drop and relative permeability data for Asmari–Jahrom and Sarkhun reservoir limestone core were measured at 140 °F and 194 °F with two types of fluids before and after being treated with methanol. Methanol treatment was found to reduce pressure drop and increase gas relative permeability. Fig. 16shows the two-phase pressure drop before and after the methanol treatment of Sarkhun limestone core at a flow rate of 5 cm³/min. The use of methanol results in lower pressure drop and consequently higher gas relative permeability. Fig. 17 compares gas relative permeability before and after methanol treatment. The treatment improved gas and condensate relative permeability by a factor of 1.6.

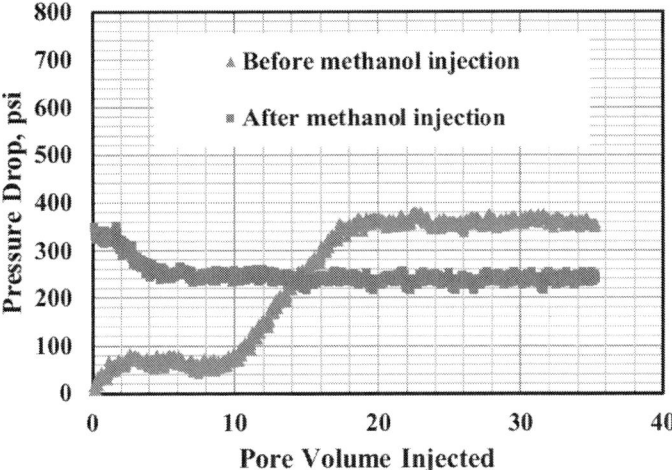

Figure 16: Pressure drop across Sarkhun cores during condensate accumulation before and after Methanol treatment at 194 °F, 1500 psig and 5 cm³/min.

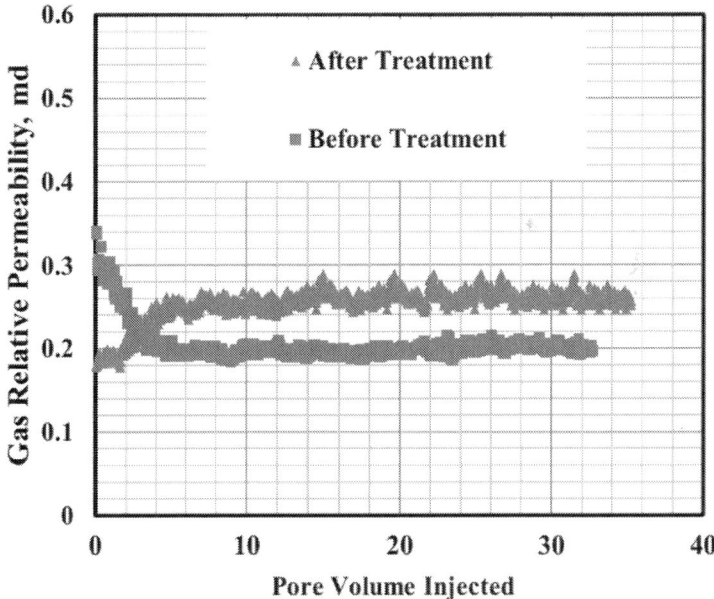

Figure 17: Gas relative permeability before and after methanol treatment at 194 °F, 1500 psig and 5 cm³/min.

Fig. 18 shows the increase in gas relative permeability in the Asmari–Jahrom limestone core resulting from methanol injection for several core floods at different initial water saturation levels. The increase in gas relative permeability varied from a factor of 1.12 to 1.64 depending on initial water saturation. Higher initial water saturation resulted in a greater increase in gas relative permeability, indicating that a dual effect of condensate and water banking is at play, thus reducing gas relative permeability. Moreover, the removal of water and condensate phases from the core by means of methanol resulted in a rise in gas relative permeability.

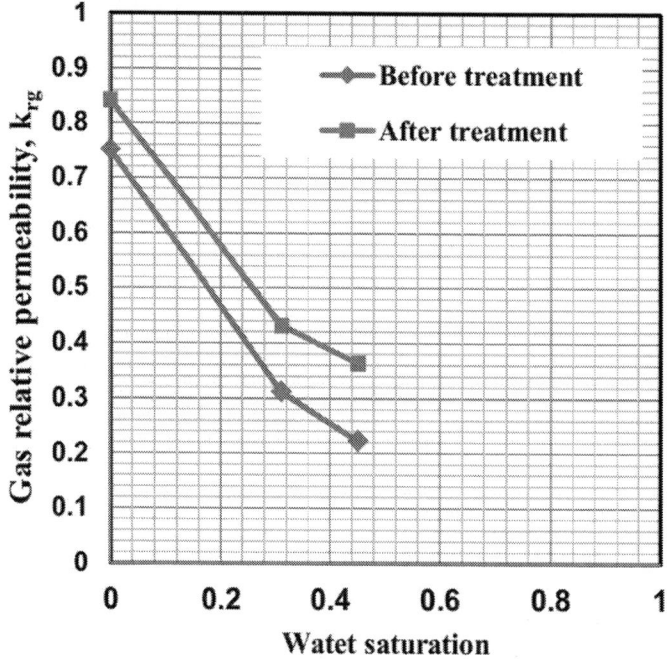

Figure 18: Effect of initial water saturation on gas relative permeability for Fluid 2 for Asmari–Jahrom limestone core.

The efficacy of the model was verified via comparison with the results that were obtained through core flood experiments. Good agreement was achieved between the experimental results and those obtained through modeling in the prediction of gas relative permeability. Table 5 shows a comparison between the model and experimental results in one of our case studies.

Table 5: Comparison between model and experimental results

	Model calculation	Experimental result	MSE	Minimum error percentage	Maximum error percentage
PCore before treatment	430.21	360.91	4274	18.27	20.13
PCore after treatment	265.12	239.91	5121	9.15	10.82
Krg before treatment	0.166	0.201	0.0011	15.21	17.32
Krg after treatment	0.231	0.262	0.0009	10.12	12.2

CONCLUSIONS

Steady state relative permeability data for Sarkhun limestone were measured at different temperatures and pressures with gas condensate fluids. The core flood experiments were carried out using the pseudo-pressure method. In this method, single-phase gas mixture with a pressure above its dew point is flashed in the upstream backpressure regulator as it enters the core at a pressure below the dew point pressure. An attractive, yet not commonly applied, calculation method called the orthogonal collocation method for solving reservoir nonlinear equations was applied in the calculations. In addition, the cubic-plus-association (CPA) equation of state, which is the most effective equation of state for the prediction of the phase behavior of alcohol–hydrocarbon mixtures, was applied in the modeling. Based on the results of this study, the following main conclusions can be drawn:

In this research, a high correlation was achieved for methanol and hydrocarbon mixtures using the CPA equation of state and the orthogonal collocation method.

For a particular fluid, both experimental work and modeling show that significant productivity loss can occur in retrograde gas condensate reservoirs when the flowing bottom whole pressure falls below the dew point pressure. Reductions of up to 80% in gas relative permeability were seen in cores due to condensate blockage.

In order to predict the impact of methanol treatment on reservoirs, a modeling procedure using the CPA equation of state and orthogonal collocation method was applied to a gas condensate fluid through the core. Gas relative permeability was found to increase by a factor of 1.4–1.6.

In addition to the modeling, the application of methanol to cores was investigated experimentally and an improvement in gas relative permeability was seen. The results show that gas relative permeability increases to approximately 30–60%.

A greater reduction in gas relative permeability occurs in the presence of water saturation. Higher initial water saturation results in a greater increase in gas relative permeability.

Increases in gas relative permeability vary from a factor of 1.12 to 1.64 depending on initial water saturation.

A good agreement was achieved between the experimental results and modeling in the prediction of gas relative permeability before and after methanol treatment.

ACKNOWLEDGMENTS

We would like to thank Dr. A. Shariati for his help with this research and we would also like to express our gratitude to the sponsors of this gas-condensate research: South Pars Oil and Gas Company.

REFERENCES

1.   Abel et al., 1970 W. Abel, R.F. Jackson, R.A. Wattenbarger Simulation of a partial pressure maintenance gas cycling project with a compositional model J. Pet. Technol. (1970), pp. 38–46

2.   Ahmed, 2001 T. Ahmed Reservoir Engineering Handbook (second ed.)Gulf Publishing Company, Houston, TX (2001), pp. 321–472

3.   Ahmed et al., 2000 H. Ahmed, E.L. Banbi, A.M. Aly, W.J. Lee, W.D. McCain Investigation of waterflooding and gas cycling for developing a gas condensate reservoir, SPE 59772 (2000)

4.   Al-Anazi et al., 2002 H.A. Al-Anazi, G.A. Pope, M.M. Sharma Laboratory measurement of condensate blocking and treatment

for the low and high permeability rocks, The SPE Annual Technical Conference and Exhibition (2002)

5. Al-Anazi et al., 2003 H.A. Al-Anazi, J.G. Walker, G.A. Pope, M.M. Sharma, D.F. Hackney A successful methanol treatment in a gas-condensate reservoir: field application, SPE 80901 (2003)

6. Al-Anazi et al., 2005 H.A. Al-Anazi, G.A. Pope, M.M. Sharma Laboratory measurement of condensate blocking and treatment for both low and high permeability rocks, SPE 77546 (2005)

7. Al-Hashim and Hashmi, 2000 H.S. Al-Hashim, S.S. Hashmi Long term performance of hydraulically fractured layered rich gas condensate reservoir, SPE 64774 (2000)

8. Alzate et al., 2006 G.A. Alzate, C.A. Franco, A. Restrepo, D.L. Barreto-Alvarez, J.J. Del Pino-Castrillon, A.A. Escobar-Murillo Evaluation of alcohol-based treatments for condensate banking removal, SPE 98359 (2006)

9. Bang et al., 2010 V. Bang, G.A. Pope, M.M. Sharma, J.R. Baran Jr., M. Ahmadi A new solution to restore productivity of gas wells with condensate and water blocks, SPE-116711-PA (2010)

10. Bonyadi et al., 2012 M. Bonyadi, M.R. Rahimpour, F. Esmaeilzadeh A new fast technique for calculation of gas condensate well productivity by using pseudo pressure method J. Nat. Gas Sci. Eng., 4 (2012), pp. 35–43

11. BP, 2013 BP BP Statistical Review of World Energy World Wide Web Address (June 2013) ⟨http://www.bp.com/⟩

12. Chen et al., 2004 J. Chen, G. Hirasaki, M. Flaum Study of wettability alteration from NMR: effect of OBM on wettability and NMR responses, Eighth International Symposium on Reservoir Wettability (2004)

13. Cullick et al., 1993 A.S. Cullick, H.S. Lu, L.G. Jones, M.F. Cohen, J.P. Watson WAG may improve gas condensate recovery, SPE 19114 (1993)

14. Danesh, 1998 A. Danesh PVT and Phase Behavior of Petroluem Reservoir Fluids Elsevier Science B. V., Netherlands (1998), pp. 142–143

15. Du et al., 2000 L. Du, J.G. Walker, G.A. Pope, M.M. Sharma, P. Wang Use of solvents to improve the productivity of gas condensate wells, SPE 62935 (2000)

16. Eilerts et al., 1965 C.K. Eilerts, E.F. Sumner, N.L. Potts Integration of partial differential equation for transient radial flow of gas condensate fluids in porous structures SPEJ (1965), pp. 141–152

17. Elliott et al., 1990 J.R. Elliott, S.J. Suresh, M.D. Donohue A simple equation of state for non-spherical and associating molecules Ind. Eng. Chem. Res., 29 (1990), pp. 1476–1485

18. Fetkovich, 1973 M.D. Fetkovich The isochronal testing of oil wells, SPE 4259 (1973) Fevang and Whitson, 1995 Ø. Fevang, C.H. Whitson Modeling gas-condensate well deliverability, SPE 30714 (1995)

19. Fishlock and Probert, 1996 T.P. Fishlock, C.J. Probert Waterflooding of gas-condensate reservoirs, SPE 35370 (1996)

20. Gilani et al., 2011 S.F. Gilani, M.M. Sharma, D. Torres, M. Ahmadi, G.A. Pope, H. Linnemeyer Correlating wettability alteration with changes in relative permeability of gas condensate reservoirs, SPE 141419 (2011)

21. Haung and Radosz, 1990 S. Haung, M. Radosz Equation of state for small, large, polydisperse, and associating molecules Ind. Eng. Chem. Res., 29 (1990), pp. 2284–2294

22. Henderson et al., 1991 G.D. Henderson, A. Danesh, J.M. Peden Waterflooding of gas condensate fluids in cores above and below the dewpoint, SPE 22636 (1991)

23. Hoier et al., 2004 L. Hoier, N. Cheng, C.H. Whitson Miscible gas injection in under saturated gas–oil systems, SPE 90379 (2004)

24. Jadhunandan and Morrow, 1991 P.P. Jadhunandan, N.R. Morrow Effect of wettability on waterflood recovery for crude-oil/brine/rock systems, SPE 22597 (1991)

25. Jones et al., 1993 L.G. Jones, A.S. Cullick, M.F. Cohen WAG process promises improved recovery in cycling gas condensate reservoirs: Part 1—Prototype reservoir simulation studies, SPE 19113 (1993)

26. Kumar, 2000 R. Kumar Productivity Improvement of Gas-Condensate Wells by Fracturing The University of Texas at Austin, Austin, TX (2000) MS Thesis

27. Kniazeff and Naville, 1965 V.Y. Kniazeff, S.A. Naville Two-phase flow of volatile hydrocarbon SPEJ, Trans. AIME, 234 (1965), pp. 37–44

28. Kontogeorgis et al., 1996 G.M. Kontogeorgis, E.P. Voutsas, I.V. Yakoumis, D.P. Tassios An equation of state for associating fluids Ind. Eng. Chem. Res., 35 (1996), pp. 4310–4318

29. Kontogeorgis et al., 1999 G.M. Kontogeorgis, I.V. Yakoumis, H. Meijer, E.M. Hendriks, T. Moorwood Multicomponent phase equilibrium calculations for water–methanol–alkane mixtures Fluid Phase Equilib., 158–160 (1999), pp. 201–209

30. Kossack and Opdal, 1986 C.A. Kossack, S.T. Opdal Recovery of condensate from a heterogeneous reservoir by injection of a slug of methane followed by nitrogen, SPE 18265 (1986)

31. Li and Firoozabadi, 2000a K. Li, A. Firoozabadi Phenomenological modeling of critical condensate saturation and relative permeabilities in gas/condensate systems, SPE 56014 (2000)

32. Li and Firoozabadi, 2000b K. Li, A. Firoozabadi Experimental study of wettability alteration to preferential gas-wetting in porous media and its effects (2000) 62515-PA

33. Lolon et al., 2003 E.P. Lolon, D.A. McVay, S.K. Schubarth Effect of fracture conductivity on effective fracture length, SPE 84311 (2003)

34. Mohan, 2005 J. Mohan Modeling of Gas Condensate Wells With and Without Hydraulic Fractures The University of Texas at Austin, Austin, TX (2005) MS Thesis   Mott, 2003 R. Mott Engineering calculation of gas condensate well productivity SPE Reservoir. Eval. Eng. J. (2003), pp. 298–306 86298

35. Muskat, 1949 M. Muskat Physical Principles of Oil Production McGraw Hill Book Company, Inc, New York, NY (1949)

36. Narayanaswamy, 1998 G. Narayanaswamy Well Productivity of Gas Condensate Reservoir The University of Texas at Austin, Austin, TX (1998) MS Thesis

37. Narayanaswamy, 1999 G. Narayanaswamy, G.A. Pope, M.M. Sharma Effect of heterogeneity on the non-Darcy flow coefficient, SPE 39979 (1999)

38. Owolabi and Watson, 1993 O.O. Owolabi, R.W. Watson Effects of rock-pore characteristics on oil recovery at breakthrough and ultimate oil recovery in water-wet sandstones, SPE 26935 (1993)

39. Rice and Do, 1995 R.G. Rice, D.D. Do Applied Mathematics and Modeling for Chemical Engineers John Wiley & Sons. Inc, USA (1995), pp. 277–287

40. Sänger and Hagoort, 1998 P. Sänger, J. Hagoort Recovery of gas condensate by nitrogen injection compared with methane injection SPEJ (1998), pp. 26–33

41. Settari et al., 1996 A. Settari, R.C. Bachman, K. Hovem, S.G. Paulson Productivity of fractured gas condensate wells—a case study of Smorbukk field, SPE 35604 (1996)

42. Smith et al., 2005 J.M. Smith, H.C. Van Ness, M.M. Abbott Introduction to Chemical Engineering Thermodynamics (Seventh ed)McGraw-Hill, New York (2005)　Solms et al., 2004 N.

43. Solms, M.L. von Michelsen, G.M. Kontogeorgis Applying association theories to polar fluids Ind. Eng. Chem. Res., 43 (2004), pp. 1803–1806

44. Walker, 2000 J.G. Walker Laboratory Evaluation of Alcohols and Surfactants to Increase Production from Gas-Condensate Reservoir The University of Texas at Austin, Austin, TX (2000) MS Thesis

45. Young, 1910 S. Young The vapor pressures, specific volumes, heat of vaporization, and critical constants of 300 pure organic substances Proc. R. Dublin Soc., 21 (1910), p. 374

46. Yakoumis, G.M. Kontogeorgis, E.C. Voutsas, D.P. Tassios Vapour–liquid equilibria for alcohol/hydrocarbon systems using the CPA equation of state Fluid Phase Equilib., 130 (1997), pp. 31–47

47. Yakoumis et al., 1998 I.V. Yakoumis, G.M. Kontogeorgis, E.C. Voutsas, E.M. Hendriks, D.P. Tassios Prediction of phase equilibria in binary aqueous systems containing alkanes, cycloalkanes, and alkenes with the cubic plus association equation of state Ind. Eng. Chem. Res., 37 (1998),

48. Zendehboudi et al., 2012 S. Zendehboudi, M.A. Ahmadi, L. James, I. Chatzis Prediction of condensate-to-gas ratio for retrograde gas condensate reservoirs using artificial neural network with particle swarm optimization Energy Fuels, 26 (6) (2012), pp. 3432–3447

GeoSys.Chem: Estimate of Reservoir Fluid Characteristics as First Step in Geochemical Modeling of Geothermal Systems

Mahendra P. Verma,

Geotermia, Instituto de Investigaciones Eléctricas, Reforma 113, Col. Palmira, Cuernavaca, Mor., C.P. 62490, México

ABSTRACT

A computer code GeoSys.Chem for the calculation of deep geothermal reservoir fluid characteristics from the measured physical–chemical parameters of separated water and condensed vapor samples obtained from drilled wells is presented. It was written as a dynamic link library (DLL) in Visual Basic in Visual Studio 2010 (VB.NET). Using this library a demonstration program GeoChem was developed in VB.NET, which accepts the input data file in the XML format.

A stepwise calculation of deep reservoir fluid characteristics of 11 production wells of Los Azufres geothermal system is performed. The calculated concentration of CO_2 (e.g.=1270 mmole/kg in the well AZ-09) in the vapor, discharged into the atmosphere at the weir box, from the water sample indicates some problem in the analysis of carbonic species concentrations. In the absence of good quality analysis of carbonic species it is suggested to consider the CO_2 in the vapor sample at the separator and the total dissolved carbonic species concentration in the water sample (i.e., without considering the liberation of CO_2 in the atmospheric vapor at the weir box) for the geothermal reservoir fluid composition calculations. Similarly, it presents various diagrams developed in Excel for the thermodynamic evolution of Los Azufres geothermal reservoir.

INTRODUCTION

Geochemical modeling of geothermal system contemplates the determination of its thermodynamic equilibrium state to enlighten the physical and chemical processes responsible for its origin and evolution (Chatterjee, 1991 and Verma, 2002). According to the second law of thermodynamics, the total entropy of the system and its surrounding is maximum in the equilibrium state. For a closed system, this state is also characterized by a minimum in its Gibbs free energy, if temperature and pressure are the independent thermodynamic variables (Garrels and Christ, 1965, Smith and van Ness, 1975 and Heidemann, 1978). It is further simplified in the form of relations between the equilibrium constants of chemical reactions and the activities (or molar concentrations in dilute solution) of chemical species, which exist in the system at the given temperature and pressure (Morel, 1983 and Chatterjee, 1991).

The computer programs for chemical modeling of the equilibrium state of multi-component fluids are useful tools for understanding water chemistry in nature as well as in the laboratory, and tracing the reaction mechanisms and processes for water-bodies evolution (Nordstrom et al., 1979, Plummer et al., 1988, Bethke, 1992 and Verma, 2002). SOLMNEQ (Kharaka and Barnes, 1973), MINEQ (Westall et al., 1976), WATEQX (van Gaans, 1989) and EQ3NR (Wolery, 1983) deal chemical speciation using input parameters as dissolved species concentration, temperature and pH; while WATEQ (Truesdell and Jones, 1974), WATCH (Arnórsson et al., 1982), CHILLER (Reed, 1982) and EQQYAC (Barragan and Nieva, 1989) may recalculate the pH using charge balance or H^+ mass-balance. NETPATH (Plummer et al., 1991) and "The Geochemist's Workbench" (Bethke, 1992 and Bethke, 1994) can also take into account mixing, dilution and evaporation processes. Nordstrom et al. (1979) reviewed over 30 chemical modeling programs and concluded that every modeling program had been developed for specific purposes with its own individual capacities and limitations. Fundamental limitations were the form of alkalinity input and non-carbonic alkalinity correction, and pH calculation. These limitations are still needed to be resolved in the revised versions of computer-programs (Verma, 2000 and Verma and Truesdell, 2001).

The first step in the geochemical modeling of geothermal system is the calculation of deep reservoir fluid characteristics from the measured physical-chemical parameters of surface manifestations like spring, fumarole and drilled well. Using the chemical geothermometers and the conservation of mass, energy and alkalinity, the chemical concentrations are converted to the reservoir conditions in order to predict the state of water-rock interaction and reservoir processes like boiling, condensation, mixing with other fluids, mineral dissolution-precipitation, etc. (Verma, 2002). The calculation procedures with certain limitations like the calculation of pH and alkalinity are systematically documented by Henley et al. (1984). Verma (2008a) developed an algorithm to calculate the reservoir fluid pH in geothermal systems with uncertainty propagation. Similarly, the quartz geothermometry is extended to estimate deep reservoir temperature and vapor fraction with multivariate uncertainty propagation (Verma, 2008b and Verma, 2012).

This article presents a computer code GeoSys.Chem, written in the object oriented programming (OOP) approach, for the first step of geochemical modeling of geothermal systems (i.e., the calculation of reservoir fluid compositions). The program is written as a dynamic link library (DLL) in Visual Basic in Visual Studio (VB.NET). Additionally, a demonstration program GeoChem is also written in VB.NET to illustrate the calculation procedure considering the Los Azufres geothermal field as an example.

THEORETICAL ASPECTS

Fig. 1 shows the conceptual diagram of geothermal system (modified after Verma, in press). The geothermal fluid flows up in the well and is separated into vapor and liquid in the separator. The vapor sample (Vapor$_2$) is collected at the separator and the separated liquid (Water$_2$) is further flushed in the weir box to collect the water sample (Water$_1$) at the atmospheric conditions. The first step of geochemical modeling of geothermal system is the reconstruction of reservoir fluid characteristics (i.e., pressure (P), temperature (T), fraction of liquid (yl) and vapor (yv), and concentration of each chemical species in the liquid (Water$_3$) and vapor (Vapor$_3$) phases).

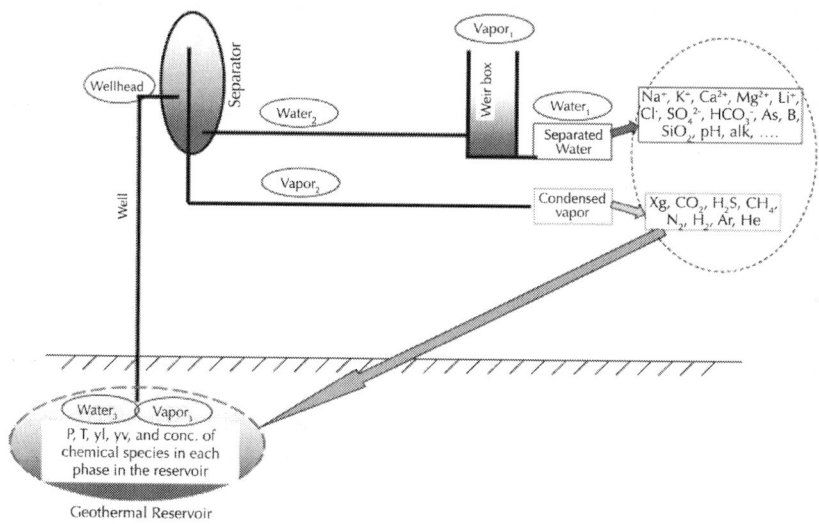

Figure 1: Schematic diagram of geothermal system (modified after Verma, in press). The vapor sample (Vapor$_2$) is collected at the separator and the separated liquid (Water$_2$) is further flushed in the weir box to collect the liquid sample (Water$_1$). The first step in geochemical modeling is the calculation of composition of the vapor (Vapor$_3$) and liquid (Water$_3$) phases in the geothermal reservoir.

The algorithm contemplates the separation of total discharge fluid into vapor and liquid at a given pressure (or temperature) along the liquid-vapor saturation curve and vice versa (Henley et al., 1984 and

Verma, 2002). Assuming adiabatic steam separation (i.e., heat loss or gain by the fluid from its surroundings is negligible) the distribution of reservoir fluid enthalpy between the liquid and vapor phases is expressed by

$$H_{td} = (1-y)H_l + yH_v \qquad (1)$$

where H is enthalpy, y is the fraction of vapor by weight and sub-indices td, v and l represent the corresponding parameter for the total discharge, vapor and liquid, respectively. The similar equation for the concentration of any chemical species is written as

$$C_{td} = (1-y)C_l + yC_v \qquad (2)$$

The non-volatile species like Na^+, Cl^-, etc., resides only in the liquid phase (i.e., their concentration in the vapor phase is zero). However, the gaseous species like CO_2, H_2S, NH_3, N_2, CH_4, etc., distribute between liquid and vapor phases. The distribution coefficient $DCoef$ for a gaseous species is defined as the concentration ratio of the species in the vapor and liquid phases.

$$D_{Coef} = \frac{C_v}{C_l} \qquad (3)$$

There are three types of equations for an aqueous solution: *mass balance, charge balance* and *proton balance*. But out of the three equations two are independent and the third can be derived as an algebraic sum of the other two equations (Verma and Truesdell, 2001). Theoretically, a solution must be electrically neutral. It is well known that the electro-neutrality condition is rarely satisfied in analytical compositions of water as a consequence of the analytical errors. Thus the alkalinity approach (i.e., proton balance) is used here for the pH calculation of geothermal fluids (Verma, 2000).

A base-neutralizing capacity (BNC) or acid-neutralizing capacity (ANC) is the equivalent sum of all the acids or bases that can be titrated with a strong base or acid to a preselected equivalence point (Stumm and Morgan, 1981). The BNC and ANC are more commonly known as acidity and alkalinity, respectively. Both of these terms are defined for certain pertinent equivalence points (EPs) for the system. Acidity is the negative of alkalinity for the same reference EP. In the carbonic (bi-proton) systems there are three equivalence points called the H_2CO_3EP, $NaHCO_3$EP and Na_2CO_3EP. Alkalinity could be defined with respect to either EP. The geothermal fluids also have other weak acids and bases like H_4SiO_4, $B(OH)_3$, H_2S, NH_3, etc. In order to handle the calculations efficiently, the alkalinity in the geothermal fluids is defined with respect to the acid equivalence point as

$$Alk=[OH^-]-[H^+]+_{CTcar}(_{\alpha1\,car}+2_{\alpha2car})+_{CTB}(_{\alpha1B})+_{CTSi}(_{\alpha1Si})+_{CTS}(_{\alpha1S})+_{CTN}(_{\alpha1N}) \quad (4)$$

where the α's are the ionization fractions (Stumm and Morgan, 1981) and CT is the total dissolved concentration of the subscripted constituent, i.e., carbonic acid (car), boric acid (B), silicic acid (Si), hydrogen sulfide (S) and ammonia (N), respectively. In case of ammonia the α's are defined for the corresponding acid (NH_4^+). Thus the alkalinity defined here does not change upon dissolution or exsolution of CO_2 (H_2CO_3) and H_2S; but it does affect with NH_3 (Stumm and Morgan, 1981). Similarly, the addition or removal of bicarbonate (HCO_3^-), carbonate (CO_3^{2-}), silicic ($H_3SiO_4^-$), boric ($B(OH)_4^-$), sulfide (HS^-, S^{2-}), hydroxide (OH^-) minerals will increase or decrease alkalinity. The precipitation of minerals is not considered in the algorithm. The fluid may be supersaturated with respect to some minerals like quartz; but there is no sufficient time to get them precipitated during the separation of fluid into the liquid and vapor phases at the separator and weir box. However, the distribution of NH_3 between the liquid and vapor phases is handled. Therefore, the alkalinity in the vapor phase is equivalent to the concentration of NH_3 in the vapor phase. In other words, the alkalinity is distributed between the liquid and vapor phases according to the following equation

$$Al_{ktd}=(1-y)Al_{kl}+yAl_{kv} \quad (5)$$

where the alkalinity in the liquid phase (alkl) is defined according to Eq. (4) and the alkalinity in the vapor phase (alkv) is turned out to be its NH_3 concentration. The alkalinity is a conservative entity during chemical reactions in a system (Stumm and Morgan, 1981).

When the separation pressure (or temperature) is known (e.g., at the separator and weir box), the calculation of composition of the liquid and vapor phases from the total discharge fluid composition and vice versa is easy to perform at the separation conditions.

In the geothermal reservoir the total discharge composition of fluid is same as the total discharge composition of fluid at the separator; however, we do not know the reservoir pressure (or temperature) in the geothermal reservoir. Therefore, the quartz solubility geothermometers are used to estimate the reservoir temperature and vapor fraction (Verma, in press).

This algorithm is programmed in the computer code GeoSys. Chem to calculate the deep reservoir fluid composition from the vapor and liquid samples collected from a production well.

PROGRAM DESCRIPTION

GeoSys

The GeoSys was written as a dynamic link library (DLL) in Visual Basic in Visual Studio 2010 (VB.NET) using the object oriented programming (OOP) approach. A namespace GeoSys was created which will be used in future to include all the programs associated with the geochemical modeling. The namespace approach is a Microsoft's naming convention in OOP which avoids the name conflict during programming. Presently, GeoSys contains three namespaces, Chem, ThemoData and Geotherms for geochemical modeling of geothermal system, thermodynamics data and chemical geothermometers, respectively. The Geotherms namespace contains the classes associated with the quartz solubility geothermometers (Verma, in press).

GeoSys.Chem

The namespace GeoSys.Chem contains various classes as shown in Fig. 2. A class encapsulates the data and methods, and serves as blueprints for creating objects. The functionality of a class may be extended without knowing its code (Verma, 2008a, b).

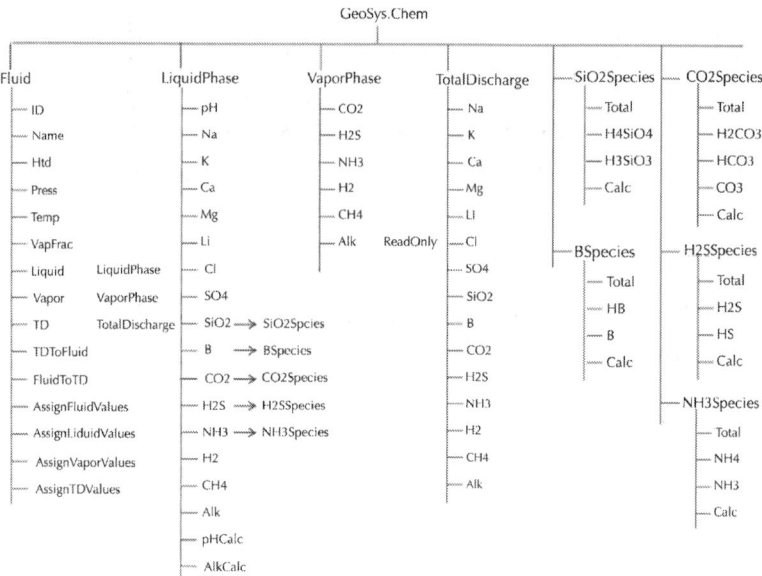

Figure 2: Classes of namespace GeoSys.Chem. Each class has properties and methods. The methods are depicted in italics. For example, the class "Fluid" has nine properties and six methods. The first two methods are the instance methods and the last four are the class methods. The user defined type properties are shown with their respective types. For example, the property, Liquid of class Fluid is of LiquidPhase type. All the properties are read and write type except the alkalinity of the class VaporPhase.

The most important class of GeoSys.Chem is "Fluid" (Fig. 2). It has nine properties (ID, Name, Htd, Press, Temp, VapFrac, Liquid, Vapor, and TD), two instance methods (TDToFluid and FluidToTD) and four class methods (AssignFluidValues, AssignLiquidValues, AssignVaporValues, and AssignTDValues). The names of properties and methods are quite explicit and are not described here. The class methods are associated with the class and are used without creating an instance of the class (i.e., object).

The properties of built-in types are shown without their types (e.g., ID as integer, Name as string, etc.); whereas the properties of user defined types are depicted with their respect types. For example, the property, Liquid of the class Fluid is of LiquidPhase type. Similarly, the properties and methods of LiquidPhase and other classes are given in Fig. 2.

GeoSys.ThermoData

The namespace GeoSys.ThermoData contains three classes: KConst, SteamTables and DCoef for the equilibrium constant for various species, thermodynamic properties of water (Verma, 2003) and distribution coefficient of gaseous species, respectively (Fig. 3).

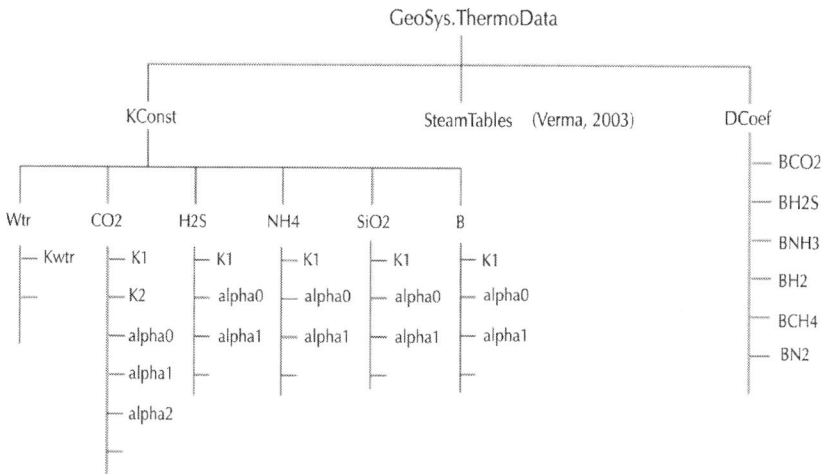

Figure 3: The namespace GeoSys.ThermoData has classes: Kconst, Steam-Tables, and DCoef for the calculation of equilibrium constants, thermodynamic properties of water, and distribution coefficients of gaseous species, respectively.

Fig. 4 shows the variation of the equilibrium constants for the species of class KConst with temperature. The values of equilibrium constants are taken from Henley et al. (1984). They reported the values up to 300 °C. Since the temperature in the geothermal reservoir is sometimes higher than 300 °C, the values are extrapolated up to the critical point of water with a linear tendency among the last three or more data points. The quadratic interpolation is carried out to obtain the intermediate values between the data points.

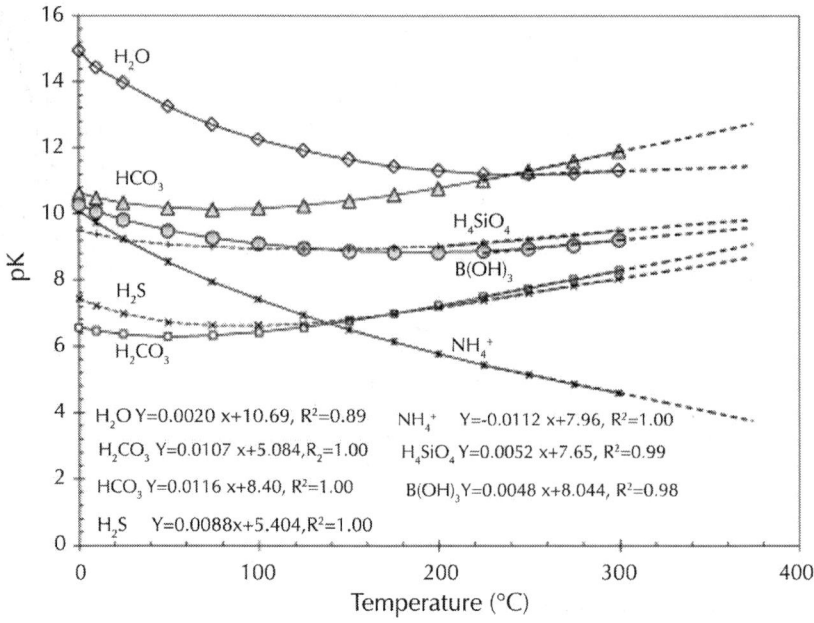

Figure 4: Variation of pK (=−log K, negative logarithm of equilibrium constant) of various species with temperature. The data were taken fromHenley et al. (1984). The values are extrapolated with linear tendency among the last three or more data points up to the critical point of water. Similarly, the intermediate values between the data points are calculated with the quadratic interpolation.

The equations for the distribution coefficients for the gaseous species are taken from Giggenbach (1980). Similarly, the equations are considered valid up to the critical point of water; although he reported the equations for the temperature range 100 to 340 °C.

GeoChem: A demo

The program GeoChem can be downloaded from the journal webpage or requested from the author. It can be installed with running the setup program and following the instructions. The default installation folder isC:\ProgramFiles\GeoSys.Chem.Fig. 5 shows its graphic user interface to calculate the deep geothermal reservoir fluid compositions. The following steps are needed to conduct these calculations:

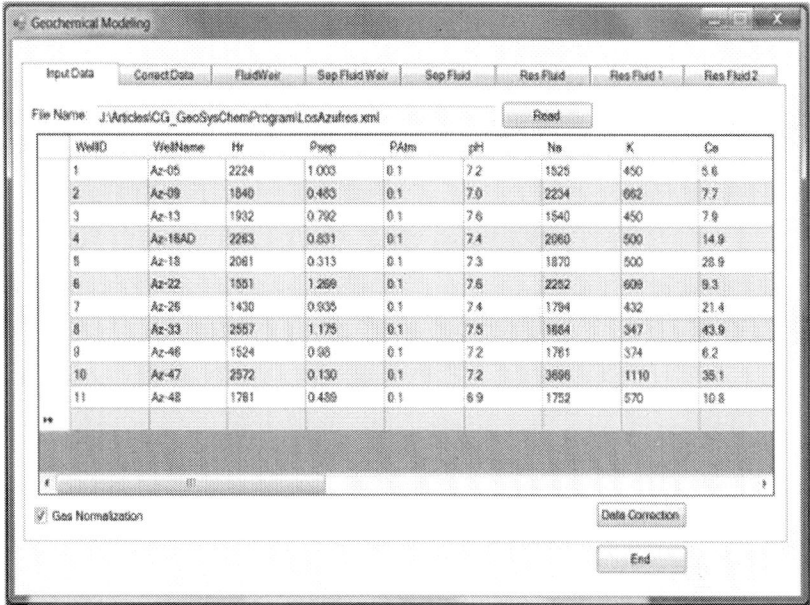

Figure 5: User interface of GeoChem.

- *Creating input data file*: The program accepts the input data file in the XML format. Table 1 shows an input data file for one well. The values are written within the opening and closing tags. For example, the total discharge enthalpy (Hr=1179 kJ/kg) is written as <Hr>1179</Hr>. The field wellID is for the identification of the well. It is planned to implement in future a database management system for the geochemical analyses.

Table 1: Input data file in the XML format for the demonstration program, GeoChem

```xml
<?xml version="1.0" encoding="utf-8" ?>
<GeoSys>
  <WellData>
    <WellID>2</WellID>
    <WellName>Az-09</WellName>
    <Hr>1840</Hr>
    <Psep>0.483</Psep>
    <PAtm>0.1</PAtm>
    <pH>7.0</pH>
    <Na>2234</Na>
    <K>662</K>
    <Ca>7.7</Ca>
    <Mg>0.05</Mg>
    <Li>0</Li>
    <SiO2>887</SiO2>
    <B>341</B>
    <Cl>4030</Cl>
    <SO4>17.7</SO4>
    <HCO3>46.6</HCO3>
    <Xg>0.285</Xg>
    <CO2>93.44</CO2>
    <H2S>4.542</H2S>
    <NH3>0.8335</NH3>
    <H2>0.7403</H2>
    <CH4>0.0902</CH4>
    <N2>0.3563</N2>
  </WellData>
</GeoSys>
```

An XML file may be created in any text editor, although many free XML editors are available on internet. The Visual Studio has its own XML editor. Open a new file in any text editor and copy and paste the content of Table 1 in it. The measurement units are: enthalpy in kJ/kg, pressure in MPa, aqueous species concentration in ppm, gas/steam ratio (X_g) in % volume, and condensed vapor species (CO_2, H_2S, ...) in % volume in the dry gas.

One can write his data values within the corresponding tags (Miller and Lawrence Berkely Laboratory Report LBL, 1979). Similarly, to add another well one has to copy the content between the tags <WellData>...</Welldata> and paste it within the tags <Geosys>...</Geosys> and save the file with an extension. xml.

- *Reading data file*: Run the program GeoChem and press the button "Read" on the tab "Input Data" (Fig. 5) and select the data file. To open the same data file every time when the program starts, save the data file in the program folder with the name, WellData.xml. It is the default file name.

- *Concentration correction*: It is necessary to convert the concentration in mmole/kg before conducting any calculation. Similarly, the normalization of gaseous species concentration is the default option. Unselect the option "Gas Normalization", if you do not want to normalize gas concentrations. Press the button "Data Correction". It will show the calculated data in the tab "Correct Data". It also calculates the charge unbalance and alkalinity for each sample. First, it calculates all the existing species. For example, the boron concentration is reported as the total concentration. Actually, there are $B(OH)_3$ and $B(OH)_4^-$ species in the solution (sample). The proportion of their concentrations depends on the pH of the solution. So, the concentration of $B(OH)_4^-$ is calculated and included in the charge unbalance calculations.

- *Reservoir fluid composition calculation*: On pressing the button "Calc" on the tab "Correct Data", it calculates the fluid compositions at various places including in the geothermal reservoir. It will be explained later in case of well AZ-09.

- *Output results*: The results are shown in the DataGridView controls. One can copy and paste them in any file.

A CASE STUDY: LOS AZUFRES GEOTHERMAL FIELD

The geochemical data of 11 production wells from the Los Azufres geothermal field are taken from Arellano et al. (2005). These data in an XML file (WellData.xml) are also uploaded on the journal webpage. Arellano et al. (2005) presented the thermodynamic evolution of the

Los Azufres geothermal reservoir using a well simulation, WELFLO (Miller, 1979). The geochemical studies also provide similar evidences on the thermodynamic evolution of Los Azufres geothermal system.

Fig. 6 shows the stepwise calculation of geothermal reservoir parameters for well AZ-09. It shows only the carbonic species due to space limits. It is helpful to install and run the program GeoChem with the default file, which contains the data of Los Azufres geothermal field. The calculation procedure is performed according to the following steps:

- *Heating liquid sample up to the weir box separation temperature*: The water samples are analyzed at the laboratory temperature (say 25 °C). First it calculates the charge unbalance to verify the analysis quality (i.e., including all the major chemical parameters). The charge unbalance for these analyses is less than 5% (see Tab "Correct Data" of the program GeoChem). It means that the analytical quality is satisfactory. The concentrations of all the carbonic species are calculated from pH and HCO_3^- concentration, measured in the water sample. Similarly, the alkalinity at 25 °C is calculated from pH and acid-base species for each sample. The well AZ-09 has the alkalinity 0.971 meq/kg and it is a conservative entity during heating the sample from 25 to 100 °C (Stumm and Morgan, 1981). Similarly, the total concentration of carbonic species (0.939 mmole/kg) is conserved, but the distribution of carbonic species and pH change on heating the sample from 25 to 100 °C. See the results on the tab "Fluid Weir" of the program.

- *Calculation of vapor composition at the weir box*: The gaseous species are liberated in the vapor phase during the liquid–vapor separation at the weir box. The carbonic species are only measured in the water sample. So, the concentration of CO_2 in the vapor phase at the weir box is calculated using the distribution coefficient (Giggenbach, 1980). The vapor has 1270 mmole/kg of CO_2. See the results for all the wells on the tab "Fluid Weir". This is a huge amount of CO_2. If it is true, the geothermal system must be in the list of most contaminating energy systems. Thus, either the measurement of HCO_3^- is incorrect or there is no liquid–vapor equilibrium in the weir box or the distribution equation for CO_2 is incorrect. Verma (2004) analyzed the dispersion among the interlaboratory comparison of HCO_3^- analyses in geothermal waters. He concluded that the analytical method used in the literature for HCO_3^- analysis is conceptually incorrect. So, there is need to improve first the bicarbonate analysis in the geothermal water samples.

- *Integration of vapor–liquid to calculate the separated water composition at the separator:* The chemical composition of separated water ($Water_{2Weir}$) at the separator is calculated by combing the analysis of water sample ($Water_1$) and the composition of liberated vapor ($Vapor_1$) at the weir box. The results for all the wells are on the tab "Sep Fluid Weir" of the program.

- *Calculation of separated water composition from vapor phase:* The gaseous species are analyzed in the vapor sample ($Vapor_2$). By considering the concentration of non-volatile species and alkalinity of the separated water ($Water_{2Weir}$) one can construct the separated water compositions ($Water_2$) from $Vapor_2$. It can be observed that there is tremendous difference in the carbonic species compositions of $Water_2$ and $Water_{2Weir}$. This separated water ($Water_2$) is flushed in the weir box at the atmospheric conditions. This also shows that there is no sufficient CO_2 in the separated water to liberate the tremendous amount of CO_2 (1270 mmole/kg) in the atmospheric vapor at the weir box (see step 2). The complete results are on the tab "Sep Fluid".

- *Calculation of geothermal reservoir fluid compositions:* The compositions of geothermal reservoir fluid are calculated by combining the separated water and vapor compositions at the separator. Three types of separated waters are considered: (i) $Water_2$, (ii) water which liberates the vapor only (i.e., without any gaseous species) at the weir box in forming the $Water_1$, and (iii) $Water_{2Weir}$. The calculated geothermal reservoir fluid compositions are given on the tabs "Res Fluid 1", "Res Fluid 2", and "Res Fluid 3", respectively. The steps 5a and 5c are the extreme cases of the process. In the absence of good quality analysis of carbonic species, the geothermal reservoir fluid compositions are considered as obtained by the step 5b (i.e., the results on the tab "Res Fluid 2").

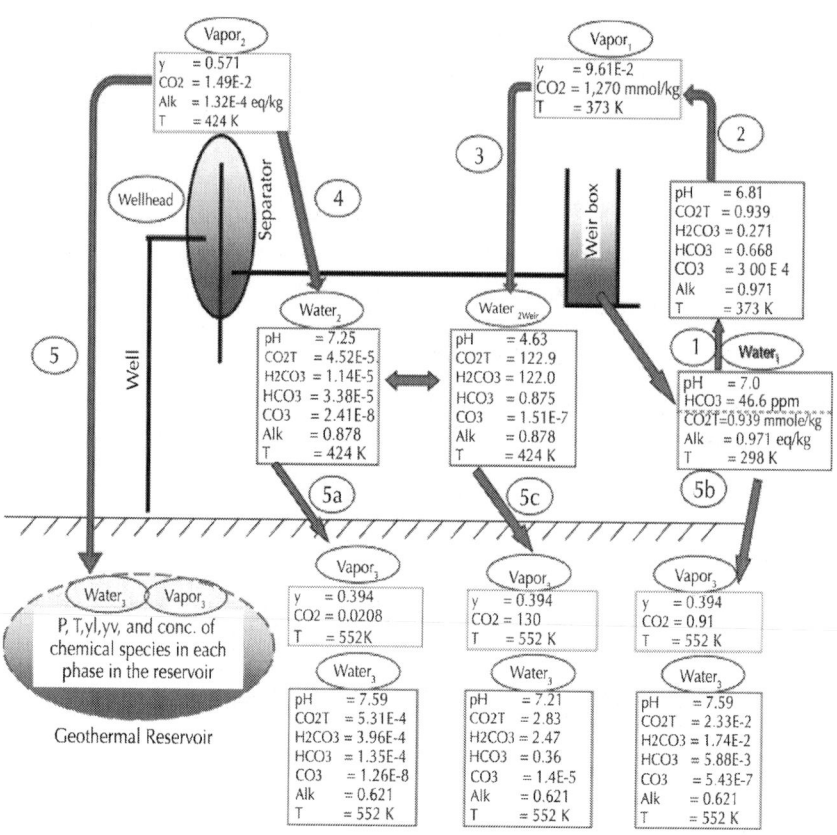

Figure 6: Illustration of geochemical calculation for well AZ-09. It shows mainly the carbonic species. The concentrations are in mmole/kg unless specified. The details of the calculation are explained in the text.

Enthalpy–Pressure Diagram

The enthalpy–pressure diagram for water is constructed in Excel using the SteamTables class (Verma, 2003). The separation boundary between the liquid and vapor phases is formed by the critical isochor and the two phase region (Fig. 7). The conditions of the geothermal reservoir fluids for all the wells are in the two phase region and between the isotherms at 200 and 300 °C. Therefore, the geochemistry of Los Azufres provides the similar evidences as obtained by the well simulator (Arellano et al., 2005).

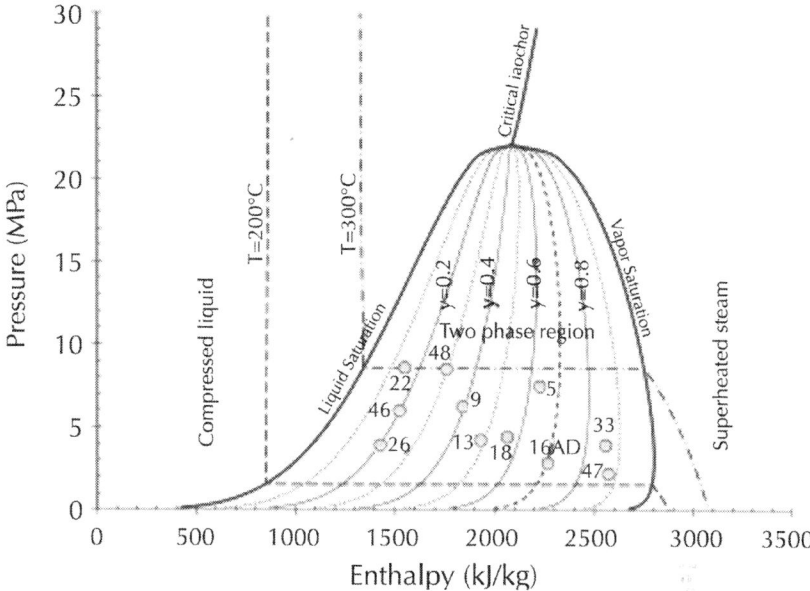

Figure 7: Enthalpy verses pressure diagram. It shows the characteristics of Los Azufres geothermal reservoir.

Cl⁻–Enthalpy Mixing Diagram

Henley et al. (1984) presented a Cl–enthalpy mixing model to evaluate reservoir processes in the formation of natural manifestations. Fig. 8 shows the total discharge and liquid phase compositions of the production wells. The corresponding vapor enthalpies for all the wells are in the range 2749–2803 kJ/kg (i.e., averaged value as 2785±21 kJ/kg). It is clear that the vapor enthalpies fall at the same point for a wide range of temperature and pressure. Similarly, the total discharge and the corresponding liquid phase values have no relation unless the fraction of vapor is known. In other words, it is not possible to predict the relation among wells if the fraction of vapor is unknown.

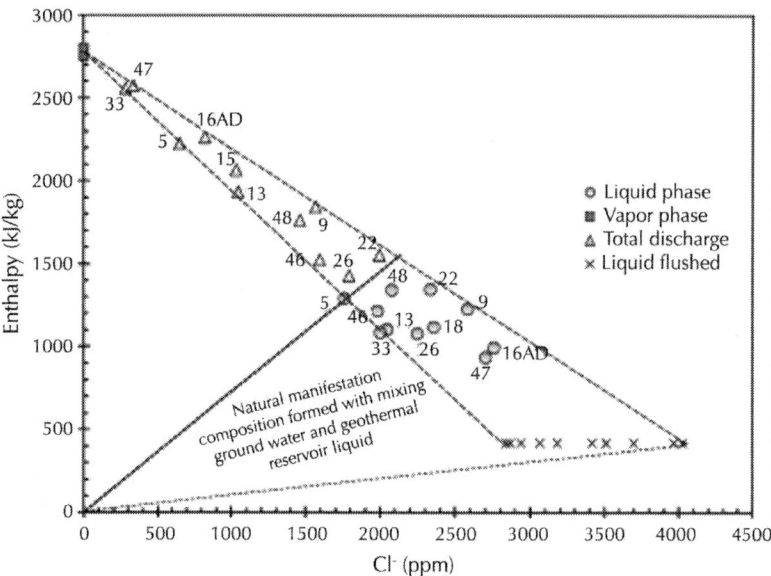

Figure 8: Chloride verses enthalpy mixing diagram for the production wells of Los Azufres geothermal system.

In the formation of natural manifestation the geothermal reservoir liquid is flushed at the atmospheric conditions. The values corresponding to each well are shown with cross marks (Fig. 8). The region for natural manifestations which is formed with the mixing of ground water and a component of geothermal reservoir is shown in the Fig. 8. It is clear that the Cl⁻–enthalpy mixing diagram works well if the formation of natural manifestations is controlled only by the geothermal reservoir liquid phase.

Behavior with Elevation

Fig. 9 shows the behavior of enthalpy, temperature, vapor fraction and Cl⁻ concentration with the elevation of production zone for each well. The production zone elevation data are taken from Nieva et al. (1987). The wells at higher elevation have higher enthalpy (Fig. 9a) and vapor fraction (Fig. 9c); whereas the temperature (Fig. 9b) and chloride (Fig. 9d) have contrary behavior. This supports the formation of geothermal production zone as a consequence of evaporation and condensation process as proposed by Nieva et al. (1987).

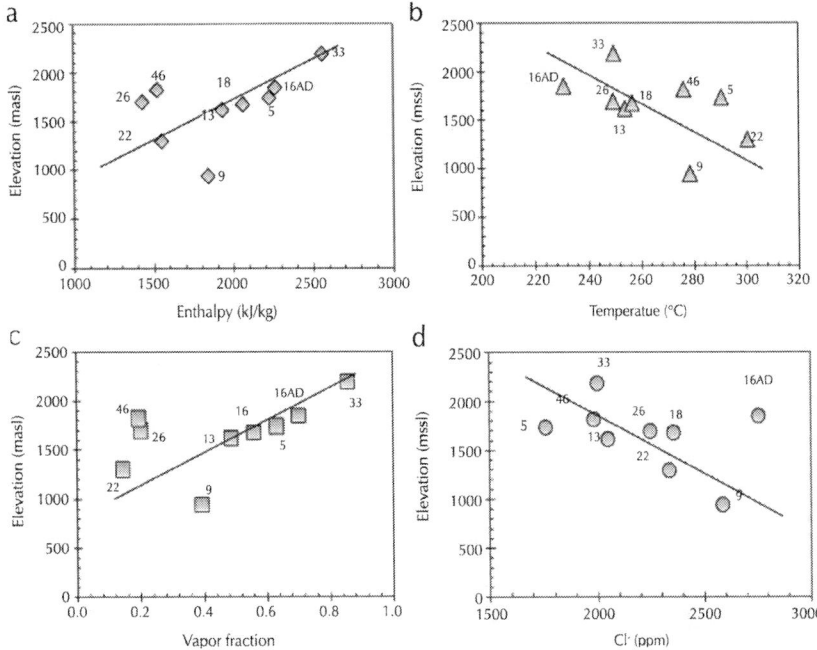

Figure 9: Variation of Los Azufres geothermal reservoir fluid parameters with the elevation of the production zone for each well.

CONCLUSIONS

The dynamic link library GeoSys.Chem, written in VB.NET, is a useful tool for the calculation of reservoir fluid characteristics during the exploration and exploitation of geothermal systems. It is programmed according to the OOP approach, so it can be incorporated and extended its functionality in any OOP language like C-sharp, VB.NET, C++, JAVA include VBA in Excel in the Windows environment.

The geochemical thermodynamic database is also implemented using the OOP approach and is applicable up to the critical point of water. The quadratic interpolation is carried out to obtain the intermediate values between the data points, since their behavior is in-between the linear and exponential tendency. A linear extrapolation is developed for calculating the thermodynamic values above 300 °C, since the tail of exponential behavior dataset is close to linear tendency.

The fundamental class of GeoSys.Chem is Fluid which has main properties Liquid, Vapor and TD (total Discharge) and methods TDtoFluid and FluidToTD. The calculation procedure of geothermal reservoir parameters is to assign the values of the properties and invoke the respective method. The vast difference in the calculated concentrations of CO2 in the secondary vapor, discharged in the atmosphere at the weir box, from the vapor sample at the separator and water sample at the weir box is a consequence of high analytical errors. Thus there is necessary to revise the analytical procedures for the determination of carbonic species concentration in the vapor and water samples (Verma, 2004).

The Los Azufres geothermal reservoir temperature is 200–300 °C and dominating process in the upper part of the reservoir is the evaporation and partial condensation. The Cl−–enthalpy diagram devised with considering the liquid component of geothermal reservoir fluid is useful for understanding the process of natural manifestations formation.

ACKNOWLEDGMENTS

This paper is dedicated to a friend and colleague, Ignacio Salvador Torres Alvarado, who passed away on 15th January, 2012. He worked actively on the geochemistry and geology of geothermal systems till the last day of his life. He will always be remembered for his friendship and helping attitude. The author highly appreciates Richard Glover and the anonymous reviewer for critical reading and providing constructive comments and suggestions to improve of the manuscript.

REFERENCES

1.   Arellano, V.M, Barragan, R.M, Torres, M.A., 2005. Thermodynamic evolution of the Los Azufres, Mexico, geothermal reservoir from 1982 to 2002. Geothermics 34, 592–616.

2.   Arno´ rsson, S., Sigurdsson, S., Svavarsson, H., 1982. The chemistry of geothermal water in Iceland I. Calculation of aqueous speciation from 0 to 370 1C. Geochimca et Cosmochimca Acta 4, 1513–1532.

3. Barragan, R.M., Nieva, D., 1989. EQQYAC: program for determining geothermal reservoir chemical equilibrium. Computers & Geosciences 15, 1221–1240.

4. Bethke, C.M., 1992. The question of uniqueness in geochemical modeling. Geochimca et Cosmochimca Acta 56, 4315–4320.

5. Bethke, C.M., 1994. The Geochemist's Workbench, Version 2.0, a User's Guide to Rxn, Act2, Tact, and Gtplot Hydrology Program. University of Illinois, USA.

6. Chatterjee, N.D., 1991. Applied Mineralogical Thermodynamics. Springer-Verlag, New York.

7. Garrels, R.M., Christ, C.H., 1965. Solutions, Minerals and Equilibria. Harper & Row Publ. Co, New York.

8. Giggenbach, W.F., 1980. Geothermal gas equilibria. Geochimica et Cosmochimca Acta 44, 2021–2032.

9. Heidemann, R.A., 1978. Non-uniqueness in phase and reaction equilibrium computations. Chemical Engineering Science 33, 1517–1528.

10. Henley, R.W., Truesdell, A.H., Barton, P.B., Whitney, J.A., 1984. Fluid-mineral equilibria in hydrothermal systems. Society of Economic Geologists, El Paso, TX.

11. Kharaka, Y.K., Barnes, I., 1973. SOLMNEQ: Solution-Mineral Equilibrium computations. NTIS Technical Report PB 214–899, 82.

12. Miller, C.W., 1979. Numerical model of transient two-phase flow in a wellbore. Lawrence Berkely Laboratory Report LBL 9056, 31.

13. Morel, F.M.M., 1983. Principle of Aquatic Chemistry. John Wiley & Sons, New York.

14. Nieva, D., Verma, M., Santoyo, E., Barraga´n, R.M., Portugal, E., 1987. Chemical and isotopic evidence of steam upflow and partial condensation in Los Azufres reservoir. In: Proceedings of the Twelfth Workshop on Geothermal Reservoir Engineering. Stanford University, Stanford, CA, pp. 253–260..

15. Nordstrom, D.K., Plummer, L.N., Wigley, T.M.L., Wolery, T.L., Ball, J.W., Jenne, E.A., Bassett, R.L., Crerar, D.A., Florence, T.M., Fritz, B., Hoffman, M., Holdren, G.R., Lafon, G.M., Mattigod, S.V, McDuff, R.E., Morel, F., Reddy, M.M., Sposito, G., Thrailkill,

J, 1979. A comparison of computerized chemical models for equilibrium calculations in aqueous systems. In: Jenne, E.A. (Ed.), Chemical Modeling of Aqueous systems 93, pp. 857–892, American Chemical Society Symposium Series.

16. Plummer, L.N., Parkhurst, D.L., Fleming, G.W., Dunkle, S.A., 1988. PHRQPITZ—a computer program incorporating Pitzer's equations for calculation of geochemical reactions in Brines. USGS Water-Resources Investigations, Report, 88–4153.

17. Plummer, L.N., Prestemon, E.C., Parkhurst, D.L., 1991. An interactive code (NETPATH) for modeling NET geochemical reactions along a flow PATH. USGS Report 91-4078, 227.

18. Reed, M.H., 1982. Calculation of multicomponent chemical equilibria and reaction processes in systems involving minerals, gases and an aqueous phase. Geochimica et Cosmochimica Acta 46, 513–528.

19. Smith, J.M., van Ness, H.C., 1975. Introduction to Chemical Thermodynamics. McGraw-Hill Kogakusha, ltd, Tokyo.

20. Stumm, W., Morgan, J.J., 1981. Aquatic Chemistry: an Introduction Emphasizing Chemical Equilibria in Natural Waters. Wiley, New York.

21. Truesdell, A.H., Jones, B.F., 1974. WATEQ a computer program for calculating chemical equilibria of natural waters. US Geological Survey and a Journal Research 2, 233–248.

22. van Gaans, P.F.M., 1989. WATEQX: a restructured, generalized, and extended FORTRAN 77 computer code and database format for the WATEQ aqueous chemical model for element speciation and mineral saturation, for use on personal computers and mainframes. Computers & Geosciences 15, 843–887.

23. Verma, M.P., 2000. pH calculation through the use of alkalinity in modelling of hydrothermal systems In: Proceedings of the Twenty-fifth Workshop on Geothermal Reservoir Engineering. Stanford University, Stanford, CA pp. 166–170.

24. Verma, M.P., 2002. Geochemical techniques in geothermal development. In: Chandrasekharam, D., Bundschuh, J. (Eds.), Geothermal Energy Resources for Developing Countries. The Swets & Zeitlinger Publishers, Netherlands, pp. 225–251.

25. Verma, M.P., 2003. Steam Tables for pure water as an ActiveX component in Visual Basic 6.0. Computer & Geosciences 29, 1155–1163.

26. Verma, M.P., 2004. A revised analytical method for HCO_3^- and CO_3^{2-} determinations in geothermal waters: an assessment of IAGC and IAEA interlaboratory comparisons. Geostandards and Geoanalytical Research 28, 1–19.

27. Verma, M.P., 2008a. IAGC and IAEA interlaboratory comparisons of geothermal water chemistry: the propagation of errors in the reservoir pH calculation. Geostandards and Geoanalytical Research 32, 317–330.

28. Verma, M.P., 2008b. QrtzGeotherm: an AcitveX component for the quartz solubility geothermometer. Computers & Geosciences 34, 1918–1925.

29. Verma, M.P. QrtzGeotherm: a revised algorithm for quartz solubility geothermometry to estimate geothermal reservoir temperature and vapor fraction with multivariate analytical uncertainty propagation. Computers & Geosciences, http://dx.doi.org/10.1016/j.cageo.2012.01.008, in press.

30. Verma, M.P., Truesdell, A.H., 2001. pH calculation through the use of alkalinity in geochemical modeling of hydrothermal systems. In: Cidu, R. (Ed.), Water Rock Interaction. Balkema, Netherlands, pp. 217–220.

31. Westall, J.C., Zachary, J.L., Morel, F.M.M. (1976). MINEQL: a computer program for the calculation of chemical equilibrium composition of aqueous systems. Technical Note No. 18, Department of Civil Engineering, Massachusetts Institute of Technology. Cambridge, M.A.

32. Wolery, T.J. 1983. EQ3NR a computer program for geochemical aqueous speciation-solubility calculations: user's guide and documentation. Report UCRL53414, Lawrence Livermore Notational Lab.

6

A 3-D Water/Rock Chemical Interaction Model for Prediction of HDR/HWR Geothermal Reservoir Performance

Zhenzi Jing[a], Kimio Watanabe[b], onathan Willis-Richards[c], and Toshiyuki Hashida[a]

[a]Fracture research institute,school of engineering,Tohoku university,sendai 980-8579,japan

[b]Richstone Ltd,TIME 24 blge 4f 2-45 Aomi,Koto-ku,Tokyo135-0064,japan

[c]Loeb Aron &Company Ltd Georgian House,63 Coleman Street,Londan EC2R 5BB,UK

ABSTRACT

A three-dimensional (3-D) water/rock chemical interaction model has been developed to examine the effect of water/rock chemical interaction (WRCI) on the long-term performance of hot dry rock and hot wet rock (HDR/HWR) reservoirs. The model, which integrates many field observations and thus generates a fracture network very similar to the natural fracture distribution in the reservoir, can predict the influence of WRCI on the overall fractured reservoir. Factors affecting WRCI and the effect of WRCI on long-term performance of Hijiori deep reservoir (Japan) have been modelled. Simulated results show that fluid chemistry, initial rock temperature, magnitude of flow rate and well spacing have a major effect on WRCI, and for such a multi-well Hijiori geothermal system, WRCI seems to make the flow distribution tend towards uniformity. The model described deals solely with chemical interactions as a function of flow rate and temperature, and takes no account of aperture variation as a result of thermoelastic effects. It is only a partial model, though it could form an important module of a coupled model.

INTRODUCTION

The hot dry rock and hot wet rock (HDR/HWR) geothermal systems involve circulating fluid through artificial and/or natural fracture networks in 'hot' rock where it is heated, and extracting the heated fluid from production well(s). The used fluid can be recirculated to move more heat from the HDR/HWR. The operation of such systems is expected to proceed for a long time, usually more than 20 years. Water/rock chemical interaction (WRCI) inevitably exerts an influence on the permeability of these high temperature and pressure circulation systems (Parker, 1997). For such a long-term circulation system, modelling is the best tool available to predict the influence of WRCI on future performance of the HDR/HWR reservoir. Unfortunately, the chemical aspects of most HDR/HWR models are poorly developed, due to a lack of adequate data related to reaction rates, an uncertainty as to the importance of ion exchange reactions, and known inadequacies in describing the microgeometry of chemical precipitation and dissolution (Willis-Richard & Wallroth, 1995 and Hayashi, Willis-Richards, Hopkirk et al., 1999). To date, only 1-D parallel plate fracture models for silica dissolution/deposition have been used to analyze the influence of WRCI (Robinson & Pendergrass, 1989, Shoji, Watanabe & Takahashi, 1990 and Jing, Willis-Richards, Watanabe & Hashida, 1996). Moreover, these studies concentrate only on assessing the physico-chemical changes in a single fracture aperture, and fail to show how WRCI exerts an influence on the overall fracture network system (Hayashi et al., 1999). Clearly, most crucial for the simulation of heat extraction is the development of a fully 3-D WRCI model. As a consequence, the aim of this paper is to incorporate WRCI into a 3-D stochastic network model. The model (Jing et al., 2000) is capable of generating a fracture network similar to that in a fractured HDR/HWR reservoir, whilst simultaneously addressing problems associated with hydraulic stimulation, fluid circulation and heat extraction. This 3-D WRCI model is the first step in developing a truly 3-D natural fracture-dominated system from earlier 1-D idealized fracture geometries in order to evaluate the influence of WRCI on fractured systems.

STOCHASTIC MODELLING OF NATURAL FRACTURES

Fracture distribution in the rock mass is a key factor in the design of an HDR/HWR reservoir. Information about orientation, dimensions, density and other mechanical properties on fracture in the rock is usually obtained from the observation of cores and well logs. Because of the complexity of fracture distribution and limited field data available, the stochastic process often represents the first approach to modelling the natural fracture system.

The key feature of this stochastic network model is that it makes full use of information on fractures obtained from borehole logs, to generate a fracture system that is very similar to the natural fracture network developed in HDR/HWR reservoirs.

The fracture network is idealized by a stochastic process with fracture properties described by probability distributions. Realizations are particular instances of the stochastic process in which the probability distributions are approximated by generated "random numbers"

In the model, the fracture network is held as a file of idealized circular fractures, each defined by position, orientation, radius and other mechanical and fracture surface properties. In principle, the model can accept fracture files from any data source or any observational stochastic data set.

Fractures are generated stochastically within a (fracture generation)

volume of $(L + 2r_{max})^3$, where L is the edge length of the model area and r_{max} is the radius of the largest fracture in the model. The fracture centers are uniformly random, following Willis-Richards et al. (1996), with the radius distribution fractal generated by equation:

$$r_\alpha = \left[(1 - \alpha) r_{min}^{-D} + \alpha r_{max}^{-D} \right]^{-1/D} \tag{1}$$

where α is a random number in the range 0 to 1, D is the fractal dimension that has been proven to be capable of mathematically representing the geometry of the natural fractures (Watanabe and Takahashi, 1995), and r_{min} and r_{max} are the specified radius of the smallest and largest fractures in the model, respectively. Using this equation, fracture $r\alpha$ can be generated by simply changing the α value. For any generated fracture, the initial fracture aperture, a_0, when evaluated at zero effective stress, is assumed to be proportional to the fracture radius, with the proportionality constant chosen to allow the undisturbed fracture network to match in situ measured permeability. This means that the fracture faces are likely to have undergone relative displacement at some time in their history, and that longer fractures are likely to have been displaced more than shorter ones. Fracture orientation distributions are based on re-sampling so as to mimic the fracture orientation distribution inferred from field observations (usually from well logs). Fractures are generated until the fracture density (m² of fracture per m³ of rock) reaches the observed level. Fig. 1 shows a typical natural fracture distribution generated by this 3-D model.

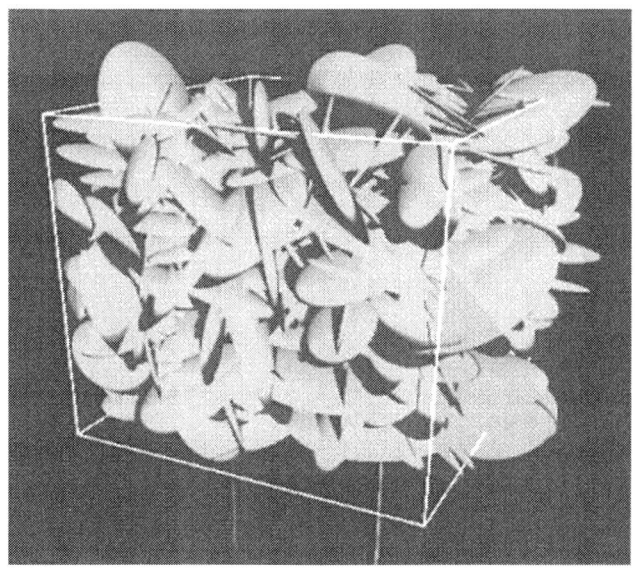

Figure 1: Image of fracture distribution generated by the 3-D stochastic network model.

WATER/ROCK CHEMICAL INTERACTION MODEL

In a closed-loop, recirculating HDR/HWR geothermal system, production fluid is reinjected into the reservoir after extracting the heat in a heat exchanger. Due to some water loss, make-up fluid needs to be added to the injection fluid, to achieve a constant production flow rate. Therefore, the injection fluid concentration, C_{in}, is controlled by:

$$C_{in} = f_m \cdot C_m + (1 - f_m)C_{out}$$

(2)

where f_m is 'make-up' fluid flow fraction, C_m is 'make-up fluid' concentration and C_{out} is production fluid concentration. The C_{out} value is governed by dissolution and deposition processes in the reservoir. A commonly used rate law for WRCI (Robinson & Pendergrass, 1989 and Watanabe, Tanifuji, Takahashi et al., 1995) in the reservoir can be expressed as:

$$\frac{\partial C}{\partial t} = \frac{S}{M} K(T)\{C^{\infty}(T) - C\}$$

(3)

where C is the concentration of dissolved rock minerals in the fluid and t is time. M is the amount of fluid involved in the reaction and S is the interfacial surface area between the rock and fluid. $K(T)$ and $C^{\infty}(T)$ are the temperature-dependent reaction rate constant and the saturation concentration.

In this study, the experimental results of $K(T)$ and $C^{\infty}(T)$ obtained on granite (Watanabe et al., 1995) were used for numerical simulation.

The fluid flow path within a fractured reservoir is approximated to a cuboid fracture with a uniform aperture, a, as shown in Fig. 2. For an infinitesimal segment, Δx, at point x with fracture aperture $a(x, t)$, and width of fracture l, then the mass balance of WRCI at time t may be expressed as

$$\rho_R \cdot \Delta x \cdot \Delta a(x, t) \cdot l = \Delta C(x, t) \cdot \rho_W \cdot Q \cdot \Delta t$$

(5)

where ρ_W and ρ_R are density of water and rock, respectively, and Q is the water flux through the fracture.

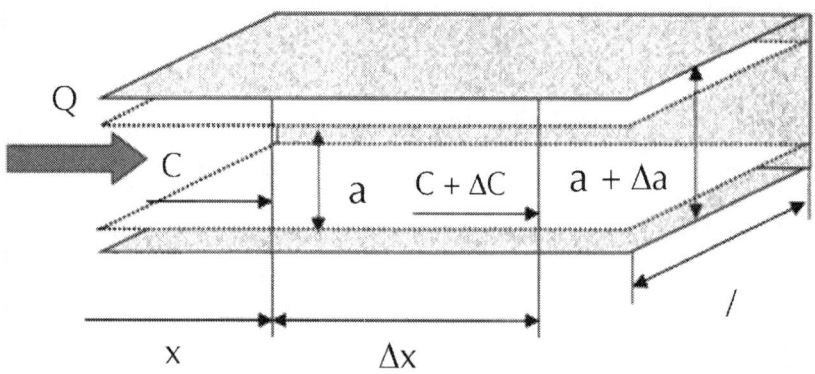

Figure 2: Simplified parallel-sided fracture model for the calculation of water/rock chemical interaction.

If $\Delta t \rightarrow 0$ and $\Delta x \rightarrow 0$, then Eq. (4) can be transformed to the following equation, which expresses the rate of aperture change along the fracture:

$$\frac{\partial a(x, t)}{\partial t} = \frac{\rho_W \cdot Q}{\rho_R \cdot l} \cdot \frac{\partial C(x, t)}{\partial x}$$

(6)

Denoting the temperature at this point by $T(x, t)$ and using Eq. (3), then the change of concentration, $\Delta C(x, t)$, is calculated as:

$$\Delta C(x, t) = \frac{2 \cdot \Delta x \cdot l}{\rho_W \cdot Q} \cdot K[T(x, t)] \cdot \{C^\infty[T(x, t)] - C(x, t)\}$$

(6)

With $\Delta x \to 0$, Eq. (6) can be expressed as the concentration change along the fracture:

$$\frac{\partial C(x, t)}{\partial x} = \frac{2 \cdot l}{\rho_W \cdot Q} \cdot K[T(x, t)] \cdot \{C^\infty[T(x, t)] - C(x, t)\}$$

(7)

Combining (5) and (7) gives:

$$\frac{\partial a(x, t)}{\partial t} = \frac{2}{\rho_R} \cdot K[T(x, t)] \cdot \{C^\infty[T(x, t)] - C(x, t)\}$$

(8)

The rate of fracture aperture change at any position in the fracture, and at any time, can be calculated using Eq. (8). The saturation concentration, C^∞, and the reaction rate constant, K, however, have to be measured before Eq. (8) may be used. The experimental data of C^∞ and K were determined by Watanabe et al. (1995) in an early study, and are used in our simulation. As previously indicated, Eq. (8) only can be used to analyze the effect of WRCI on a single fracture aperture. For a natural fracture network, the influence of WRCI on overall permeability requires further consideration, as discussed below.

In a 3-D stochastic network there are many thousands, even millions, of fractures generated in a reservoir volume of one km³. Steady state flow and heat transfer in the 3-D natural fracture system may be solved numerically by the finite differential method (FDM), with the square grid element (block) approach used in our calculation method shown in Fig. 3. For each block, some fractures will intersect with its faces [seeFig. 3(a)]. The intersected fractures on each face are assumed to be independent of each other and are approximated statistically to cuboid fractures, each of intersecting width l_i, aperture a_i and interposing fracture length L (equal to the block length), as shown in Fig. 3(b). Fluid is inferred to flow through every intersected fracture, since the fracture network is assumed to be well connected. For each intersected fracture, aperture change due to mineral dissolution or deposition can be deduced by Eq. (8). The average flow rate through each fracture is used to infer WRCI. For a given thermal history of individual block, the average aperture change for each fracture on that block face can be readily calculated, assuming that the fluid concentration in the block is known.

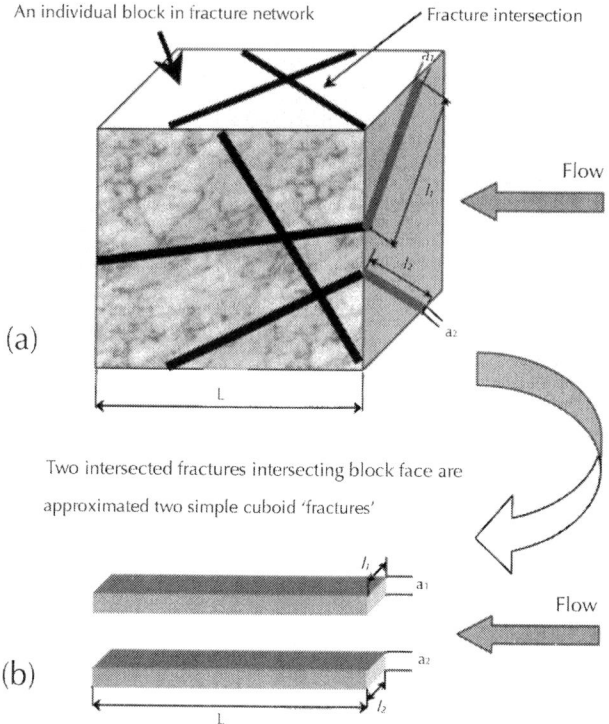

(a)

(b)

Figure 3: Concept of water/rock chemical interaction model. The fractures intersecting each block face can be approximated by rectangular fractures.

The fluid concentration for a particular block will depend on the sum of the product of the magnitude of fluid concentration and flow rate from neighbouring blocks, divided by the total flow rates flowing from those neighbouring blocks. Blocks intersected by injection well(s) are assumed to have the same fluid concentration as the injection fluid. The fluid concentration of the block (i,j), for example, in the 2-D case is shown in Fig. 4, and calculated by the following equation:

$$C_{i,j} = \frac{Q_{i-1,j}C_{i-1,j} + Q_{i,j-1}C_{i,j-1} + Q_{i+1,j}C_{i+1,j}}{Q_{i-1,j} + Q_{i,j-1} + Q_{i+1,j}}$$

(9)

where $Q_{i,j}$ and $C_{i,j}$ are the flow rate and fluid concentration for the block (i,j).

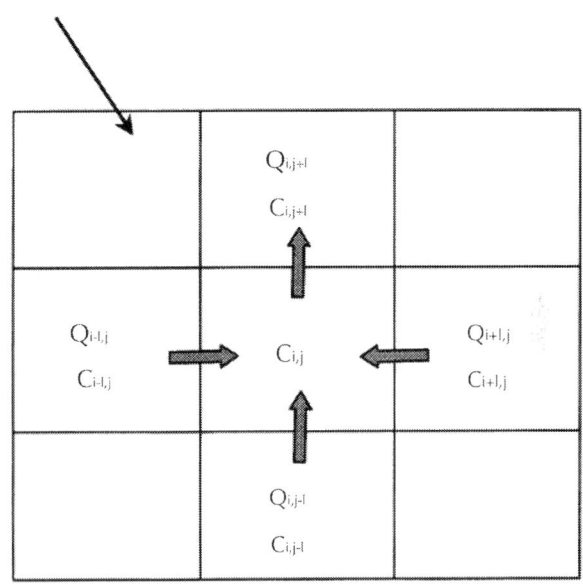

Figure 4: Two-dimensional model for calculation of fluid concentration. The calculation method only deals with the fluid flowing into the C_{ij} block (see text for explanation).

The calculated fluid concentration, C_{ij} for block (i,j), is subsequently used to calculate the fluid concentration of its neighbouring blocks. By this method, the fluid concentration of all blocks in the model can be calculated.

Similarly, the relationship between permeability (discussed in more detail in a later section) and WRCI can be assessed block by block. The steady state fluid flow and heat extraction in the network correspond closely to deduced variations in permeability. By iteration, the effect of WRCI on fluid circulation, and heat extraction in the reservoir through time, can be modelled.

As mentioned above, the experimental data for granite were used to examine the effect of the dissolution and deposition process on the reservoir performance. An experiment carried out by Wang et al. (1996) for simulating water/rock (granite) interaction in an HDR has demonstrated that the dissolution and precipitation of silica mineral were dominant among the chemical species in the granite. Furthermore, it was observed from microscopic observation of the sample surface that the weight loss of granite below 300°C was mainly due to the dissolution of feldspar and above that temperature the dissolution of quartz was dominant.

STIMULATION, CIRCULATION AND HEAT EXTRACTION

The 3-D stochastic network model has the capability of simultaneously addressing outstanding problems associated with hydraulic stimulation, fluid circulation and heat extraction and then lends confidence to the thermal predictions.

Hydraulic Stimulation

Stimulation causes shearing on pre-existing fractures, which in turn leads to a permanent increase in aperture by shear dilation. Shear displacement invariably results in some degree of shear dilation, due to the inherent roughness of fracture walls. Because the stress regime in the Earth's crust is usually in a near-critical state (Evans et al., 1999), a relatively small decrease in effective stress due to an increase in fracture fluid pressure may cause shear slip of fractures even in regions that are tectonically inactive. The change in aperture due to slip, *as* is the product of shear displacement and the tangent of the effective shear dilation angle (Goodman, 1976, Barton, Bandis & Bakhtar, 1985 and Willis-Richards, Watanabe & Takahashi, 1996):

$$a_s = U \tan\left(\phi_{dil}^{eff}\right)$$

$$(10)$$

where U is shear displacement and (ϕ_{dil}^{eff}) is effective shear dilation angle at a given normal stress.

From the sheared aperture, the fracture aperture, a, can be approximated by (Willis-Richards, Watanabe & Takahashi, 1996 and Hicks, Pine, Willis-Richards et al., 1996):

$$a = \frac{a_0 + U \tan(\phi_{dil})}{1 + 9(\sigma - p_n)/\sigma_{nref}}$$

$$(11)$$

where a_0 represents the initial (i.e. unstimulated) total compliant aperture of the fracture; n_{ref} is the effective normal stress applied to cause a 90% reduction in the compliant aperture; n is the total normal stress; p is fluid pressure; and $_{dil}$ is the dilation angle, which is obtained from. (ϕ_{dil}^{eff})

Eq. (11) provides a simple equation that can be incorporated into the model, even with up to several millions of fractures, each of which may have a small displacement.

Flow Circulation

In the 3-D model, fluid flow is assumed to be laminar, with the quantity of fluid flow from block to block controlled by Darcy's law. The discrete network model is solved by converting it to an equivalent continuum mesh. The procedure of conversion is illustrated in Fig. 5. The "permeability" contribution from each fracture is governed by the sum, for each block face, of the product of the cube of the fracture aperture (affected by WRCI) and the fracture intersection length with each block face. The steady state flow solution is solved numerically using a Multigrid algorithm (Press et al., 1992), which is very useful for quickly obtaining an approximate solution that may be refined by Gauss–Seidel simultaneous over-relation (SOR). In the fluid flow calculation, the variation in fracture aperture will respond closely to the local pressure, through Eq. (11), with further shear being discounted at this stage.

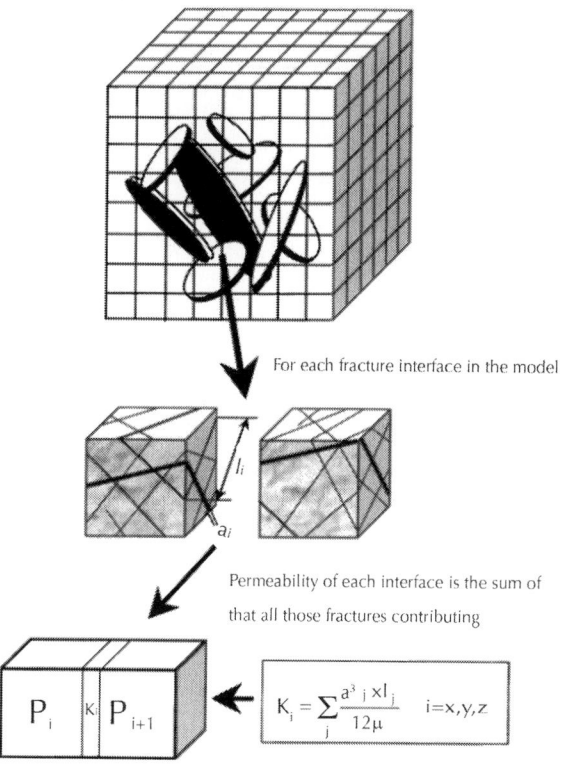

For each fracture interface in the model

Permeability of each interface is the sum of
that all those fractures contributing

$$K_i = \sum_j \frac{a^3{}_j \times l_j}{12\mu} \quad i=x,y,z$$

Figure 5: Concept of converting the discrete fractures to the equivalent permeability at the element interfaces.

Fluid loss through the boundaries of the 3-D model is estimated semi-analytically by embedding the explicitly-modelled volume within a spherical shell containing non-stimulated but naturally compliant fractures, as shown in Fig. 6. The outer boundary of the shell (at $r=\infty$) is taken as constant pressure (hydrostatic) and along its thickness the permeability is prescribed as a function of pressure. The rock mass between fractures is assumed to be impermeable, so that the fluid loss only deals with the fluid flowing to the far field. The fluid loss to the far field depends on both the pressure difference between the edge of the model area (r=boundary) and the shell's outer boundary ($r=\infty$) and the average permeability at some point between the model and outer boundaries. This approach has been used, with some success, to estimate fluid loss during circulation and for stimulation, at a number of HDR sites (Willis-Richards, 1995).

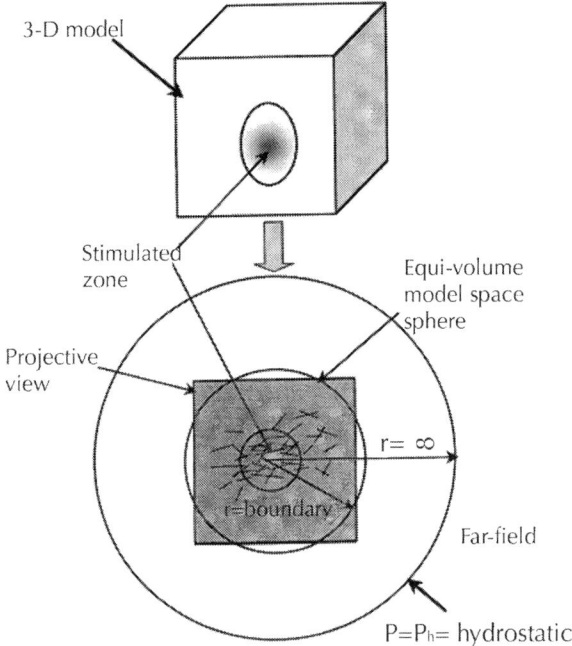

Figure 6: Idealized HDR/HWR geothermal system for the concept of fluid loss.

The model does not implement thermal stresses and temperature dependent fluid viscosity in this study.

Heat Extraction

Heat extraction is based on the same finite difference grid as the flow calculation and was solved by the alternative direction implicit method so as to model large times, with the assumption that thermal equilibrium is reached between each solid element and the water passing through the fracture (Watanabe and Takahashi, 1995). Heat is also transferred by conduction from block to block. This assumption is reasonable for moderate flow rates and moderately high fracture densities, as are typical of many HDR circulation experiments. Constant temperature boundary conditions are assumed. The calculated temperature distribution in the reservoir, at each calculation time step, is in turn used to model the effect of WRCI during circulation.

FACTORS AFFECTING WATER/ROCK CHEMICAL INTERACTION

The 3-D WRCI model has been used to examine the effects of WRCI on the long-term performance of an HDR/HWR system. The parameters used for the simulation are shown in Table 1, in which most parameters, such as fracture geometry, stress state, rock fracture properties and permeability etc., derive from actual field data from the Hijiori deep reservoir (Japan), whilst other values are assumed. The polar coordinate plots of the normals to fractures observed on BHTV images of well HDR-2a in Hijiori are shown in Fig. 7.

Table 1: Model parameters for sensitivity study and Hijiori deep reservoir simulation

Fracture geometry	
Fractal dimension of fracture radius	2.4 [e]
Fracture density (m²/m³)	0.7
Fracture orientation (dip, azimuth)	HDR-2a[a]
Stress state	
Vertical stress at center of model volume (MPa)	50
Maximum horizontal stress at center of model volume (MPa)	60[b]
Minimum horizontal stress at center of model volume (MPa)	35[b]
Vertical stress gradient (MPa/m)	−0.0261
Maximum horizontal stress gradient (MPa/m)	−0.0682[c]
Minimum horizontal stress gradient (MPa/m)	−0.0426[c]
Azimuth of maximum horizontal stress	100°[d]
Rock properties	
Young's modulus (GPa)	60
Poisson's ratio	0.25
Density (kg/m³)	2700
Specific heat (J/kg)	900
Thermal conductivity (W/m.K)	3.0

Rock fracture properties	
Fracture basic friction angle	40°
Shear dilation angle	2.5°[e]
90% closure stress (MPa)	20
Permeability	
In-situ average (m^2)	10^{-14}[c]
Fluid properties	
Density (kg/m^3)	1000
Specific heat (J/kg)	4200
Viscosity (N s/m^2)	3×10^{-4}
Fracture simulation	
Smallest fractures simulated (m)	10
Largest fracture simulated (m)	150
Temperature	
Injection fluid temperature (°C)	45
Initial rock temperature for sensitivity study (°C)	250
Initial rock temperature of Hijiori deep reservoir (°C)	240
Temperature gradient (°C/km)	40
Circulation	
Pressure (MPa)	8
Simulation	
Pressure for sensitivity study (MPa)	20
Pressure of Hijiori E9508 (MPa)	14
Well-spacing	
Sensitivity study (m)	200
Hijiori deep reservoir	
HDR1–HDR2 (m)	90
HDR1–HDR3 (m)	130

- NEDO, 1994.
- NEDO, 1995.
- NEDO, 1991.
- Tezuka, 1997.

- Jing, 1998.

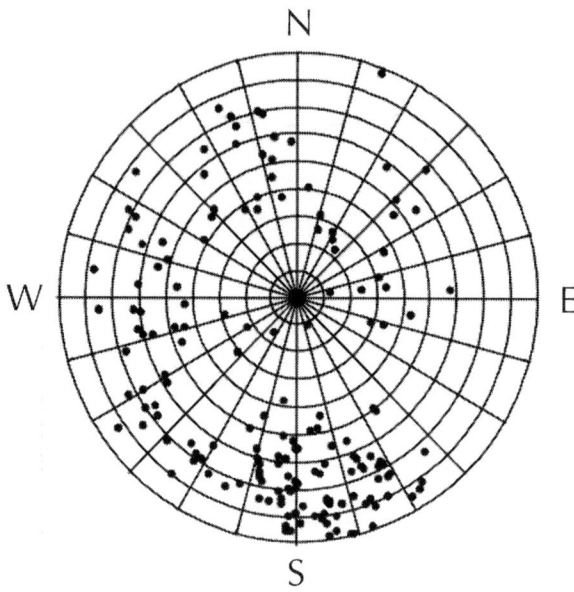

Figure 7: Pole plots of the fractures observed on a BHTV image of well HDR-2a.

The dimensions of the model are 1000×1000×1000 m, with the simulated region located at a depth of 1900 m. Sensitivity studies carried out by Jing (1998) showed that in this and similar cases an element of up to 20×20×20 m can satisfy the precision requirements for the analysis of an HDR/HWR geothermal system.

The effect of WRCI on permeability can be estimated by the change of flow rate, since mineral dissolution or deposition in the fractured medium results in a widening or narrowing of the fracture aperture and this inevitably causes a change to flow rate.

HDR/HWR reservoirs tested to date have produced fluid at the range of 5–25 kg/s. But even a small commercial geothermal power plant producing, for example, 5 MW of electricity from water at 150°C would require flow rates on the order of 150 kg/s (Abé et al., 1999). For this reason, the economic targets for an HDR/HWR reservoir›s flow rates were set around 75–100 kg/s (Garnish, 1985, Shock, 1986 and Parker, 1999). Since the magnitude of flow rate has a significant effect on changes in fracture aperture (Jing et al., 1996a) due to WRCI, the injection flow rates in this study were produced in the range of 10–63 kg/s so as to obtain a large range of information about the effect of WRCI on long-term performance of such reservoirs.

Effect of Make-up Fluid on WRCI

In a closed-loop recirculating HDR/HWR system, some amount of 'make-up fluid' (assumed to be fresh water in this study) needs to be added to the injection fluid so as to maintain operating pressures, and provide balance against the fluid loss that occurs during circulation. In such a circulating system, the concentration of dissolved rock in the injection fluid is controlled by the fraction, f_m, of make-up (fresh) fluid. For example, if f_m=1, then only make-up fluid is injected to the system. In contrast, if f_m=0, then no make-up fluid would have been added, and the net amount of rock in injection fluid removed from the production fluid would effectively be zero. Fig. 8 shows that WRCI exerts an influence on performance of HDR/HWR in such a closed-loop recirculating system. When production fluid is reinjected without removal of dissolved rock minerals (i.e. f_m=0), then injection flow enhancement with time due to WRCI is less than if fresh water (i.e. f_m=1) is injected [Fig. 8 (1)]. When make-up fluid is injected, a driving force for dissolution is produced near the injection wellbore. A driving force for inhibiting dissolution, or even pro-deposition, occurs when only production fluid is reinjected, with the resultant effect that the fracture aperture near the injection wellbore may decrease. This means that the make-up fluid has the potential to make permeability around the injection well(s) increase. If the flow rate of reinjected production fluid is large, then the rate at which the aperture decreases in size is also enhanced. A lower flow rate apparently results in little change to aperture size (Jing et al., 1996a), so reinjecting production fluid at a low flow rate should have a negligible effect on overall permeability.

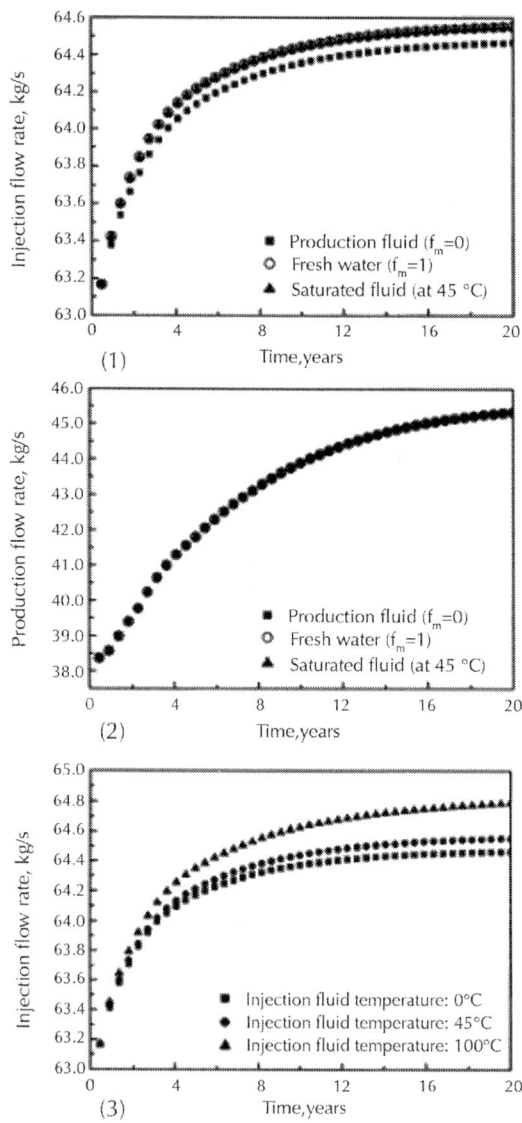

Figure 8: Effect of 'make-up fluid' on flow rate due to the influence of WRCI. The make-up fluid has little influence on production flow rate.

The saturation concentration in the injection fluid, at an injection temperature of 45 °C, is very low and there is little difference in injection flow rate enhancement by reinjecting either fresh water or the saturated concentration, as shown in Fig. 8(1 and 2). For the following simulation, the saturation concentration at the given injection temperature is assumed to be the same as the concentration of the injection fluid. The influence of WRCI on production flow rate is shown in Fig. 8(2) and, for all cases, the effect on flow rate is minimal. This means that the make-up fluid has little influence on production flow rate.

The effect of (fresh water) injection fluid temperature on WRCI is shown in Fig. 8(3). A higher temperature for injected fresh water leads to a greater influence on injection flow rate, since a higher temperature fluid has more effect on rock solubility. This behaviour suggests that pure hot water (i.e. hot water containing no dissolved rock) may be used to dissolve precipitated minerals, instead of an acidization treatment.

Effect of Initial Rock Temperature on WRCI

In Equation 3, the saturation concentration, C^∞, and reaction rate constant, K, are dependent on temperature, and initial rock temperature in the reservoir will affect WRCI. Since the first experiment in the late 1970s at Fenton Hill, New Mexico (Duchane, 1997), HDR/HWR field experiments have been carried out to exploit thermal energy from several HDR sites worldwide. Initial rock temperatures in these reservoirs were in the range of 100–270°C. Although work is under way to operate at much higher temperature, in the 400–500°C range (Hashida et al., 1998), the study on WRCI at such high temperatures is beyond our present state of knowledge. The initial rock temperatures, therefore, are assumed to be 150, 200, 250 and 300°C respectively, in this study, and their effect on flow rate at an injection pressure of 8 MPa is shown in Fig. 9. Greater injection and enhancements to production flow rate are indicated with an increase of initial rock temperature, with WRCI very sensitive to initial rock temperature in the reservoir. For example, the flow rates remain almost unchanged in the case where the initial rock temperature is 150°C, because the low initial rock temperature causes little WRCI and subsequently only a very small permeability change in the reservoir. Where, however, the rock temperature is 200, 250 or 300°C, then WRCI exerts a major effect on flow rate. It can be inferred that a greater temperature will increase the water/rock reaction rate, and thus that changes in flow rate are also more pronounced at higher temperature. This behavior means that a hotter (or deeper) reservoir will lead to a larger influence of WRCI on reservoir permeability, whilst in a lower temperature (or shallower) reservoir the influence of WRCI on permeability is limited. It is noted that, where the initial rock temperature is 300°C, the flow rate is three times greater than where the rock temperature is 250°C. This indicates that there is a need to study further the influence of WRCI at higher initial rock temperatures, even at near- or super-critical temperatures.

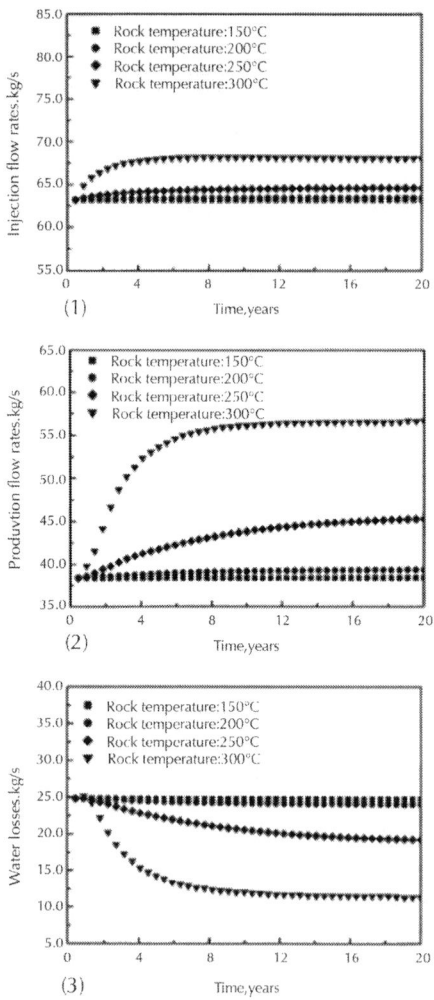

Figure 9: Effect of initial rock temperature on flow rates and water loss due to the influence of WRCI. The higher the temperature, the greater the effect. Without the influence of WRCI, the change of flow rate and water loss will be zero.

For a rock temperature of 300°C, as shown in Fig. 9(2), there is initially a rapid increase in production flow rate, which levels off after approximately six years. In contrast, a more gradual increase in production flow rate is shown over the same period for a rock temperature of 250°C. During the initial circulation period, the solubility of the rock constituents is greater, due to the higher rock temperature, which results in an increase in permeability. With temperature drawdown in the reservoir, however, the solubility of the rock components decreases and after some time the WRCI effect on permeability becomes less pronounced. The increase of production flow rate is also affected by the increase of injection flow rate, with the latter being smaller than the production flow rate [Fig. 9(1 and 2)], which results, after some time, in a leveling off of the flow out of the system. For a rock temperature of 250°C the WRCI, due to the lower rock temperature, helps to increase the production flow rate gradually. This suggests that, for a reservoir with a high initial rock temperature, the main influence on WRCI and performance of the HDR system occurs during the period of initial circulation.

The effect of WRCI on water loss is similar to that on flow rate, as shown in Fig. 9(3), i.e. a higher initial rock temperature results in less water loss, because the WRCI causes a lower impedance between the injection and production well.

Effect of WRCI on Injection and Production Flow Rates

Fig. 9(1 and 2) also shows the effect of rock temperature (and consequently WRCI) on injection and production flow rates. The influence on production flow rate is more pronounced than that on injection flow rate, for cases where initial rock temperature is 200, 250 and 300°C, respectively. The higher the initial rock temperature, the greater the influence. This phenomenon is best explained by suggesting that rocks adjacent to the injection wellbore cool rapidly, to a temperature at which mineral dissolution is low, whilst dissolution processes continue in the high temperature region near to the production wellbore. Thus, the impedance near the production wellbore reduces with time, and this leads to an increase in production flow rate.

Heat extraction from an HDR/HWR reservoir creates a thermal cooling front, which is initially located near to the injection area and progresses with time towards the production wellbore. This in turn leads to a zone(s) in the reservoir where either dissolution or deposition might occur, as the system evolves. The history of mean aperture change along fractures connecting the injection and production wells, for a given initial rock temperature of 250°C, is shown in Fig. 10. At first, the main zone of dissolution occurs near to the injection wellbore, but it moves towards the production wellbore with time. It is noted that, after 20 years, the mean aperture enhancement around the production wellbore is much larger than around the injection wellbore. This provides strong evidence to explain the greater influence of WRCI on production flow rate than on injection flow rate.

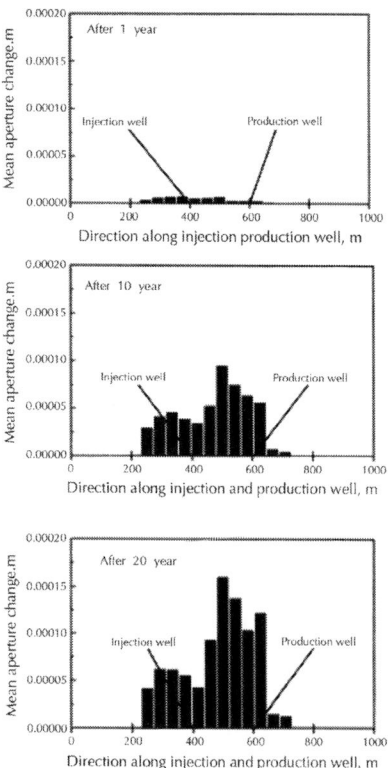

Figure 10: History of mean aperture change of fracture in vicinity of injection and production wells, as a result of WRCI.

Effect of Flow Rate on WRCI

Different injection pressures were used to produce different flow rates in the model so as to examine the effect of flow rate on WRCI. The effect of flow rate over time (and subsequently WRCI) for injection pressures of 2, 4 and 8 MPa, respectively, at initial rock temperature of 250°C, is shown in Fig. 11. A larger flow rate is produced by a greater injection pressure, which in turn causes a greater WRCI influence and flow rate enhancement. For example, for an injection pressure of 8 MPa, production flow rate increases from 38.36 to 45.33 kg/s (by 7.0 kg/s) after 20 years' circulation, whilst production flow rates increase from 16.93 to 22.10 kg/s (5.17 kg/s) and 6.38 to 8.20 kg/s (1.82 kg/s) for injection pressures of 4 and 2 MPa, respectively, over the same period. This means that the magnitude of flow rate has an obvious influence on WRCI, and the larger the flow rate, the greater the influence. This simulated result agrees with the experimental results obtained by Wang (1996). Compared to the enhancement of injection flow rate, the enhancement of production flow is much larger, especially for larger flow rates. The influence of injection pressure on water loss change, and hence on WRCI, is shown in Fig. 12. The figure shows that a greater flow rate produced by a higher injection pressure results in less water loss over time. The impedance has decreased due to WRCI, and production flow rate (during circulation), due to decreased water loss, is consequently increased.

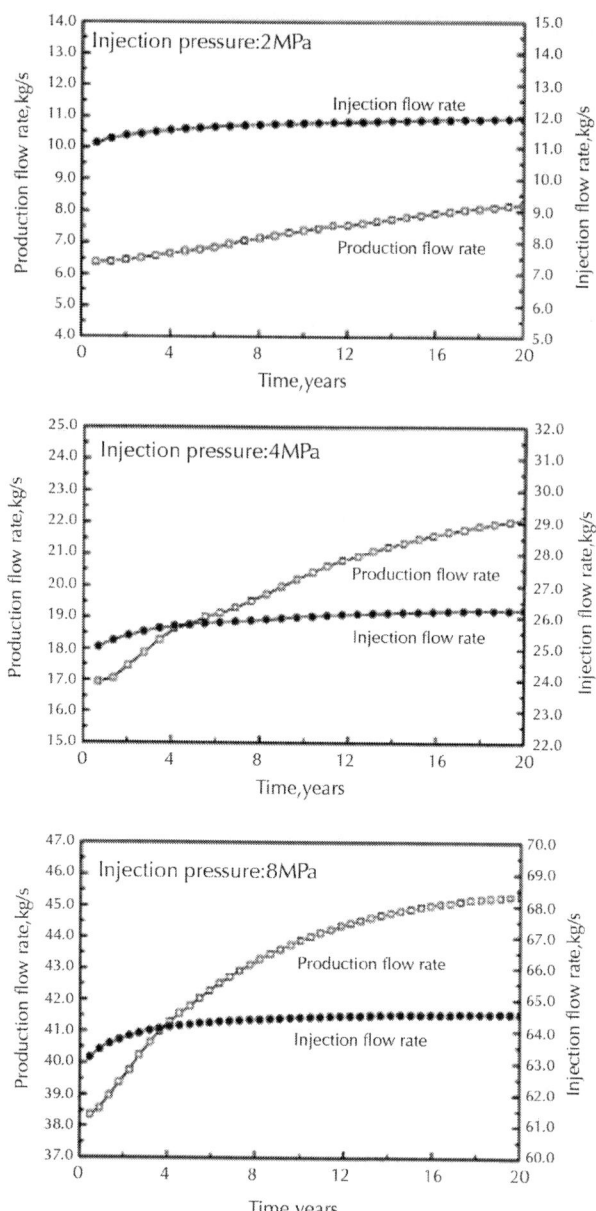

Figure 11: Effect of injection pressures on production flow rates due to WRCI. The larger flow rate produced by larger injection pressure results in a greater WRCI influence, especially for production flow rates.

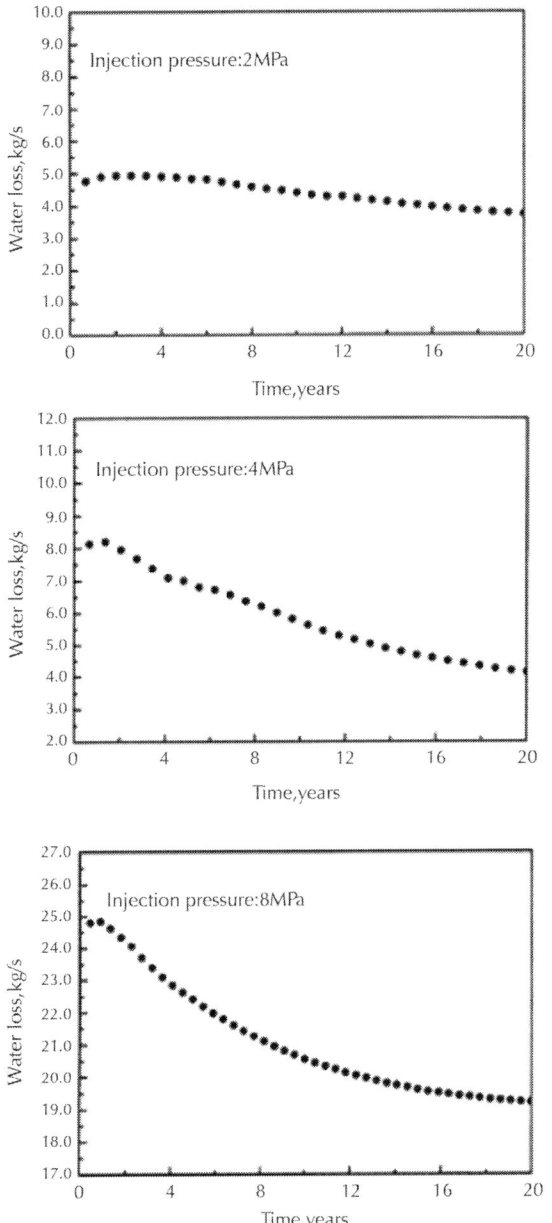

Figure 12: Effect of injection pressure on water loss with time due to the influence of WRCI. The larger flow rate produced by the higher pressure leads to a greater decrease in water loss.

Effect of Well Spacing on WRCI

Well spacing strategy differs in various HDR projects; for example, in the Ogachi geothermal (Japan) system well spacing is about 80 m, whilst in the Soultz geothermal system (France) well spacing is 450 m. It is therefore worthwhile looking at the influence of well spacing on WRCI. Fig. 13(1 and 2) shows the effect of well spacing on injection/production flow rate and production temperature due to WRCI at an injection pressure of 8 MPa and rock temperature of 250°C. The two cases presented are for a well spacing of 100 m and 200 m, respectively, for which we have assumed the same initial injection and production flow rates, so as to remove the influence of the magnitude of the flow rate. For the larger well spacing of 200 m, WRCI apparently exerts a greater influence on both injection and production flow rates, while for the smaller well spacing of 100 m this effect is less marked. We suggest in a reservoir with larger well spacing that there is more time for dissolution processes to take place (and occur more readily), before reservoir temperature decrease, as shown in Fig. 13(3), i.e. a larger well spacing also results in higher temperatures being maintained in the reservoir. We suggest that more attention should be paid to larger HDR systems, where well spacing may be shown to have a major influence on WRCI, and on the long-term performance of the system.

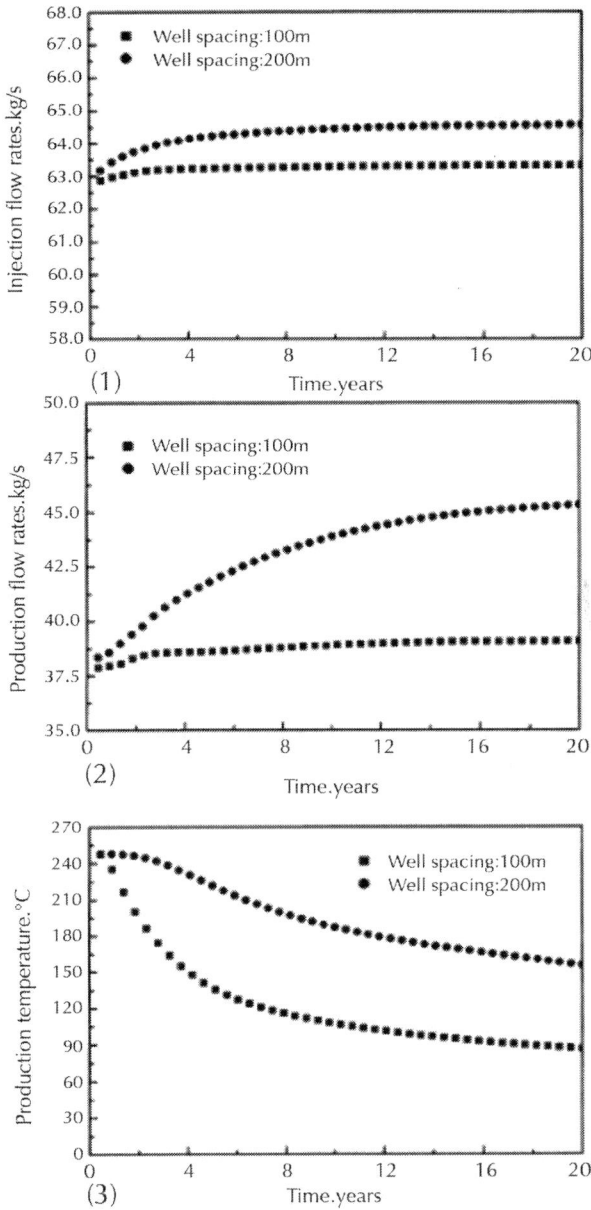

Figure 13: Effect of well spacing on flow rates and production temperature because of the influence of WRCI. For the bigger reservoir system (i.e. with larger well spacing), WRCI has a greater influence on permeability enhancement.

APPLICATION TO HIJIORI FIELD EXPERIMENT

An experimental HDR system has been developed in the Hijiori area, Japan, consisting (initially) of a shallow reservoir at a vertical depth of about 1800 m and a (subsequent) deep reservoir at a depth of about 2200 m. A 25-day preliminary circulation test was carried out in 1995, for injection well HDR-1, and production wells HDR-2a and HDR-3. Following a series of experiments, the shallow Hijiori reservoir was modelled by Willis-Richards, Watanabe & Takahashi, 1996, Kruger & Yamaguchi, 1993 and Jing, Willis-Richards, Watanabe & Hashida, 1996. Few simulations, however, have been carried out for the deep reservoir at Hijiori, especially to predict the effects of WRCI on the long-term performance of the Hijiori system. In this study, the 3-D WRCI model was used to evaluate the effects of WRCI on the long-term performance of the Hijiori deep reservoir. The data collected during creation of, and circulation within, the deep reservoir were used as input to this model so as to simulate the effect of WRCI on the Hijiori deep reservoir behaviors. The parameters of Hijiori deep reservoir are listed in Table 1.

The simulation has been undertaken to examine the influence of WRCI on a 5-year performance of the Hijiori deep reservoir system, as shown in Fig. 14(1). Simulated flow rates were 16.7, 4.0 and 2.9 kg/s, respectively, which are similar to the Hijiori experimental result: HDR-1: 16.7 kg/s, HDR-2a: 4.0 kg/s and HDR-3: 3.0 kg/s, respectively, which remain almost unchanged over the period of circulation modelling. Consequently, we infer that WRCI exerts a negligible influence on the long-term performance of the Hijiori deep reservoir.

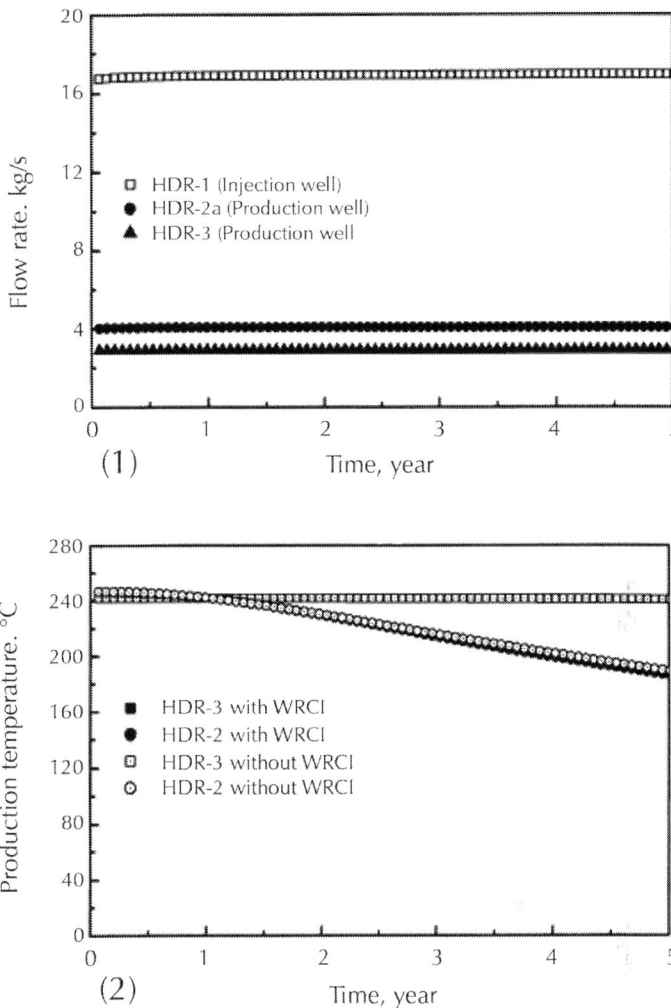

Figure 14: Effect on (1) flow rate and (2) production temperature, as a consequence of WRCI, at Hijiori deep reservoir. WRCI has little influence on the long-term performance of Hijiori deep reservoir under the condition of E9501.

In the Hijiori experiment, the injection flow rate was about 16.7 kg/s and only about 40% of the fluid was recovered from the production wells (HDR2a-3). As previously discussed, production flow rates are insufficient to cause a WRCI effect on permeability. Simulated production temperature drawdown is shown in Fig. 14(2) and indicates that both production temperatures (i.e. with or without considering WRCI) are almost the same. This means that WRCI is unlikely to exert a major effect on temperature drawdown in the Hijiori deep reservoir.

If the rock temperature is assumed to vary from the actual (240°C) Hijiori reservoir temperature to 260 and 300°C, then the simulated results (Fig. 15) show an increase in flow rates, especially for the 300°C case. This means that even under current conditions at Hijiori, there is potential for WRCI to have a major influence on the deeper (hotter) reservoir, and hence on the long-term performance of the system. It should be noted that, in the cases of a 260 and 300°C initial rock temperature, there is initially a marked increase in HDR-2 flow rate, as shown in Fig. 15(2), which later stabilizes, whilst flow rates for HDR-3 [Fig. 15(3)] increase more gradually. The reason for this phenomenon in the multi-well Hijiori deep reservoir system may be related to the different thermal history and well spacing between HDR-1/HDR-2a and HDR-1/HDR-3, which results in a permeability contrast. The temperature contrast between HDR-1 and HDR-3 is greater than between HDR-1 and HDR-2a [similar to Fig. 14(2)]; moreover, the well spacing between HDR-1 and HDR-3 is about 130 m, whereas the spacing between HDR-1 and HDR-2a is 90 m (Nagai and Tenma, 1997). As indicated, the hotter and longer flow path (i.e. from HDR-1 to HDR-3) seems to have caused more dissolution reactions to occur in the fractured medium, which in turn leads to the change of impedance between HDR-1 and HDR-3 becoming less than between HDR-1 and HDR-2. With no obvious increase in injection flow rate during this time, the decrease in flow rate from HDR-2a is inferred to directly relate to the ease at which fluid may flow through HDR-3. This suggests that, in a multi-well HDR/HWR system, WRCI has the potential to make flow distribution tend towards uniformity, i.e. with a reduced impedance to flow the permeability distribution tends towards uniformity.

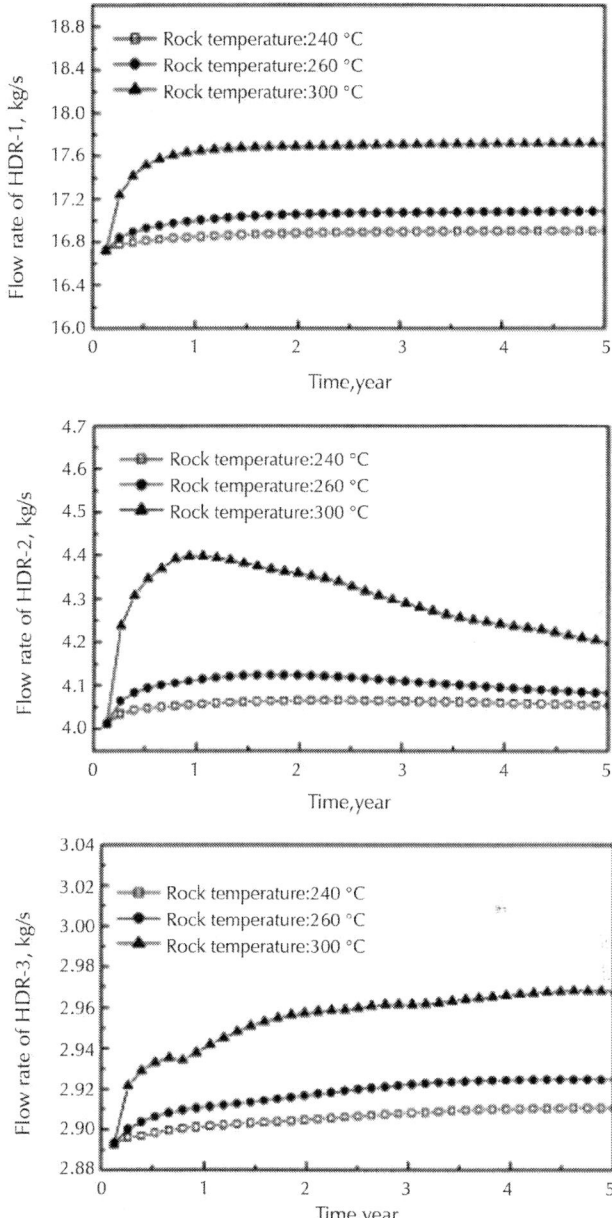

Figure 15: Effect on injection and production flow rate, as a result of WRCI, at Hijiori deep reservoir. The different flow path leads to a different flow rate enhancement with time between HDR1–HDR2 and HDR1–HDR3.

CONCLUSIONS

A 3-D water/rock chemical interaction model for an HDR/HWR reservoir has been developed to examine the influence of WRCI on overall permeability in the geothermal reservoir. The model has been used to analyze factors affecting WRCI and to predict the effects of WRCI on the long-term performance of the Hijiori deep reservoir.

In a closed-loop recirculating HDR/HWR system, if fresh fluid is injected, the potential for dissolution exists near the injection wellbore. Therefore, make-up fluid has the potential to increase the permeability around the injection well(s), and high temperature fresh water seems to be capable of dissolving precipitated rock minerals in a fractured medium. Make-up fluid only exerts an effect on injection flow rate, due to the fact that dissolution predominantly takes place near to the injection wellbore.

WRCI is very sensitive to initial rock temperature in the reservoir. The higher the initial rock temperature, the greater the influence of WRCI on permeability. This means that, for a hotter or deeper reservoir, WRCI will have a greater effect on permeability. The influence of WRCI mainly occurs during an initial circulation period, due to higher rock temperatures during this period.

WRCI exerts a stronger effect on production flow rate than on injection flow rate because rocks near to the injection wellbore quickly cool to a temperature at which dissolution is negligible.

Flow rate has an influence on WRCI, in that the larger the flow rate, the greater the water/rock chemical influence. This suggests that the influence of WRCI on permeability can be determined by the magnitude of flow rate, especially production flow rate.

For a bigger reservoir system (i.e. with larger well spacing), WRCI has a greater effect on permeability enhancement. We should therefore be paying more attention to larger HDR/HWR geothermal systems, when developing realistic models for HDR/HWR systems.

This model has been used to predict the influence of WRCI on the long-term performance of the Hijiori deep reservoir. Simulated results show that WRCI has little effect on flow rate and production drawdown. However, if the initial rock is assumed to increase to 300°C (i.e. a deeper reservoir is induced), WRCI may cause a permeability enhancement that could lead to an increased flow rate.

In the case of the multi-well Hijiori deep reservoir system, WRCI exerts a different influence on impedance for different flow paths. The hotter and longer flow path between HDR1 and HDR3 seems to induce more rock dissolution and a rapid decrease in impedance, while the cooler and shorter flow path from HDR-1 to HDR-2 leads to impendence decreasing more slowly. This behaviour suggests that WRCI has the potential to make flow distribution tend towards uniformity.

Finally, it is worth emphasising that the present model deals solely with chemical interactions, and takes no account of thermoelastic effects. The model is only a partial model, though it could form an important module of a fully coupled model. Such an investigation is now under way.

ACKNOWLEDGEMENTS

The authors wish to express their gratitude to the late Professor H. Takahashi, Tohoku University, for his encouragement and useful advice, and would like to thank Dr. G. Bignall for his helpful discussions. The work reported here was partially supported by the Japan Society for the Promotion of Science under Grant-in-Aid for Research for the Future Program (JSPS-RFTF 97P00901).

REFERENCES

1. Abe, Duchane, Parker & Kuriyagawa, 1999 H. Abé, D.V. Duchane, R.H. Parker, M. Kuriyagawa Present status and remaining problems of HDR/HWR system design Geothermics, 28 (1999), pp. 573–590

2. Barton, Bandis & Bakhtar, 1985 N. Barton, S. Bandis, K. Bakhtar Strength deformation and conductivity coupling of rock joints International Journal Rock Mechanics and Mining Sciences and Geomechanics Abstracts, 22 (3) (1985), pp. 121–140

3. Duchane, 1997 Duchane, D., 1997. Hot Dry Rock in the USA: Where is it going? Proceedings, NEDO Internal Geothermal Symposium, 311–315.

4. Evans, Cornet, Hashida et al., 1997 K.F. Evans, F.H. Cornet, T. Hashida, K. Hayashi, T. Ito, K. Matsuki, T. Wallroth Stress and rock mechanics issues of relevance to HDR/HWR engineered geothermal system: Review of developments during the past 15 years Geothermics, 28 (1999), pp. 455–474

5. Garnish, 1985 Garnish, J.D., 1985. Hot Dry Rock — a European perspective. Proceedings of the Geothermal Research Council International Symposium on Geothermal Energy, International Volume, Hawaii, 329–337.

6. Goodman, 1976 R.E. Goodman Methods of Geological Engineering in Discontinuous Rocks West, New York (1976)

7. Hashida, Takahashi, Sato & Nakatsuka, 1998 Hashida, T., Takahashi, T., Sato, K., Nakatsuka, K., 1998. Fracture Mechanics Study on Formation Process of Artificial Geothermal Reservoirs under Supercritical Water Conditions. International Conference — 4th HDR Forum, Strasbourg, France.

8. Hayashi, Willis-Richards, Hopkirk et al., 1999K. Hayashi, J. Willis-Richards, R.J. Hopkirk, U. NiiboriNumerical models of HDR geothermal reservoirs — A review of current thinking and progress Geothermics, 28 (1999), pp. 507–518

9. Hicks, Pine, Willis-Richards et al., 1996 T.W. Hicks, R.J. Pine, J. Willis-Richards, S. Xu, A.J. Jupe, N.E.V. Rodrigues A hydro-thermo-numerical model for HDR geothermal reservoir evaluation Int. J. Rock Mech. Min. Sci. and Geomech. Abstr., 33 (5) (1996), pp. 499–511

10. Jing, 1998 Jing, Z., 1998. Simulation of Heat Extraction from Fractured Geothermal Reservoir. PhD thesis, Tohoku University.

11. Jing, Willis-Richards, Watanabe & Hashida, 1996 Jing, Z., Willis-Richards, J., Watanabe, K., Hashida, T., 1996a. The effect of water/rock interaction on the permeability of HDR geothermal reservoir. Proc. 2nd International Conference on Solvothermal Reactions. Takamatsu, Kagawa, Japan, 169–172.

12. Jing, Willis-Richards, Watanabe & Hashida, 1996 Jing, Z., Willis-Richards, J., Watanabe, K., Hashida, T., 1996b. A new 3-D network model for HWR geothermal reservoir in fractured basement. 1996 Annual Meeting Geothermal Research Society of Japan — abstracts with Programs, University of Tokyo, Japan, 8–10 December, B14 (in Japanese).

13. Jing, Willis-Richards, Watanabe & Hashida, 2000 Z. Jing, J. Willis-Richards, K. Watanabe, T. Hashida A 3-D stochastic rock mechanics model of engineered geothermal systems in fractured crystalline rock Journal of Geophysical Research, 105 (B10) (2000), pp. 23663–23680

14. Kruger & Yamaguchi, 1993 Kruger, P., Yamaguchi, T., 1993. Thermal draw down analysis of Hijiori HDR 90-day circulation test. Proc. of the 18th Workshop on Geothermal Reservoir Engineering. Stanford University, Stanford, CA, 111–118.

15. Nagai & Tenma, 1997 Nagai, M., Tenma, N., 1997. Development of Hot Dry Rock technology at Hijiori Test Site- Program for a long term circulation test. Proceedings of NEDO International Geothermal Symposium, NEDO, Sendai, Japan, 11–12 March, 351–356.

16. New Energy Development Organization, 1991 New Energy Development Organization, 1991. FY 1990 Summary of Hot Dry Rock geothermal power project in Japan (in Japanese).

17. New Energy Development Organization, 1994 New Energy Development Organization, 1994. FY 1993 Summary of Hot Dry Rock geothermal power project in Japan (in Japanese).

18. New Energy Development Organization, 1995 New Energy Development Organization, 1995. FY 1994 Summary of Hot Dry Rock geothermal power project in Japan (in Japanese).

19. Parker, 1997 Parker, R., 1997. HDR academic review: Overall HDR System Design. Tohoku University, Sendai, Japan, March 14–16, pp. 1–13.

20. Parker, 1999 R. Parker The Rosemanowes HDR Project 1983–1991 Geothermics, 28 (1999), pp. 603–615

21. Press, Teukolsky, Vetterling & Flannery, 1992 W.H. Press, S.A. Teukolsky, W.T. Vetterling, B.P. Flannery Numerical Recipes in Fortran — The Art of Scientific Computing Cambridge, Cambridge University Press (1992)

22. Robinson & Pendergrass, 1989 Robinson, B.A., Pendergrass, J., 1989. A combined heat transfer and quartz dissolution-deposition model for a Hot Dry Rock geothermal reservoir. Proc. 14th Stanford Annual Workshop on Geothermal Reservoir Engineering, Stanford University, CA, 121–135.

23. Shock, 1986 Shock, R.A., 1986. An economic assessment of hot dry rocks as an energy source for the UK. Energy Technology Support Unit Report ETSU–R–34. Publ. HMSO (London), pp. 215.

24. Shoji, Watanabe & Takahashi, 1990 Shoji, T., Watanabe, K., Takahashi, H., 1990. Long term performance of geothermal circulation system. Significance of water rock interaction. Camborne School of Mines International Conference on Hot Dry Rock Geothermal Energy, pp. 436–445.

25. Tezuka, 1997 Tezuka, K., 1997. Study on the characteristics of artificial Hijiori HDR reservoir by AE method. PhD thesis, Tohoku University (in Japanese).

26. Wang, 1996 Wang, Y., 1996. Experimental study of fluid and hot dry rock interactions under geothermal conditions. PhD thesis, Tohoku University (in Japanese).

27. Wang, Yamasaki, Tsuchiya et al., 1996 Y. Wang, N. Yamasaki, N. Tsuchiya, K. Nakatsuka, S. Nishiushi Hydrothermal dissolution on granite rock in fluidized tube reactor in a temperature gradient Journal of the Geothermal Research Society Japan, 18 (4) (1996), pp. 253–262 (in Japanese)

28. Watanabe & Takashashi, 1995 K. Watanabe, H. Takahashi Fractal geometry characterization of geothermal reservoir fracture networks Journal of Geophysical Research, 100 (1995), pp. 521–528

29. Watanabe, Tanifuji, Takahashi et al., 1995 Watanabe, K., Tanifuji, K., Takahashi, H., Wang, Y., Yamasaki, N. & Nakatsuka, K. Water/rock chemical interaction under reservoir condition. Proc. of the 20th Workshop on Geothermal Reservoir Engineering 20, Stanford University, pp. 189–195.

30. Willis-Richards, 1995 J. Willis-Richards Assessment of HDR reservoir stimulation and performance using simple stochastic models

31. Geothermics, 24 (1995), pp. 385–402

32. Willis-Richard & Wallroth, 1995 J. Willis-Richards, T. Wallroth Approaches to the modelling of HDR reservoirs: a review Geothermics, 24 (1995), pp. 307–332

33. Willis-Richards, Watanabe & Takahashi, 1996 J. Willis-Richards, K. Watanabe, H. Takahashi Progress toward a stochastic rock mechanics model of engineered geothermal systems Journal of Geophysical Research, 101 (B8) (1996), pp. 17481–17496

An Automated Approach for an Optimised Least Cost Solution of Reinforced Concrete Reservoirs Using Site Parameters

A. Stanton and A.A. Javadi

Department of Engineering, University of Exeter, Exeter, UK

ABSTRACT

This paper presents design, development and application of a finite-element based least cost optimisation model (named ResOp) for reservoirs using a Genetic Algorithm. The model makes use of site specific parameters not normally considered at outline design but which are usually available; such as site plan limits, maximum height above ground level and geotechnical conditions.

The results show that such site based parameters have a significant effect on cost which can be easily incorporated at outline design stage without making expensive changes at the detailed design stage of a project. This would also be suitable when considering a selection of sites. Current cost models in the industry are too basic and should become more site specific.

The design of a reservoir constructed in Cornwall was compared to an optimised reservoir design using ResOp. The results show a potential for substantial savings to be made. The aspect ratio and shape found reasonable correlation to best practice, but the developed model suggests a more refined optimisation approach which includes site variables.

INTRODUCTION

Reinforced Concrete (RC) is extensively used due to its thermal properties and its resilience to chemical attack, particularly in underground or partially buried reservoirs. A reinforced concrete reservoir can be almost any shape or size and the storage tank can be elevated above ground, at ground level or below ground level. In the past waterbars were used extensively for RC reservoirs, but due to leakage and maintenance issues monolithic construction has been more popular. Concrete reservoirs also can have a healing process which can repair cracks that appear on the face that is in contact with water. Autogenous healing can occur for cracks up to 0.3 mm wide [20].

Although many mathematical optimisation techniques have been available in research for decades, it has only been a recent development that the latest structural design software now incorporates these more complex design refinements. As the building project lifecycle has relied more heavily upon software, and the costs and the environmental impact of civil engineering projects have been scrutinised in recent years, a trend has been found toward the optimisation of structures which can lead to cost reductions of design, construction, maintenance and demolition. This in turn reduces material wastage and material transport away from site.

Scia Engineer by Nemetschek is a commercial structural engineering graphical software system for design, calculations and verifying various codes of practice. It uses the latest technology of Object Orientated CAD conforming to buildingSMART's 'openBIM' standards. It is capable of analysing models created using other Building Information Modelling (BIM) compatible software and can use the imported objects directly in the analysis. It conforms to the latest Eurocode 2 Part 3 for the design of liquid retaining and containment structures which can design crack widths propagating from the surface of the concrete. Scia Engineer uses XML (Extensible Mark-up Language) as its main communication between third party programs and its output. The benefit of this language is that the output can easily be created in the form of a readable document.

Visual Basic for Applications (VBA) is the programming language built into all MS Office programs for its Component Object Model (COM) programming model. Excel and Scia Engineer fully support this COM programming model and therefore shall be used in this project as the link between the two programs but the code shall be executed in MS Excel.

Global optimisation is less well known in design of reinforced concrete reservoirs as the procedures are far more complex and require more computation. Scia Engineer has much documentation on optimisation and global optimisation using a Genetic Algorithm. MOOT (Multi-Objective Optimisation Tool) can adjust the size, length and properties of almost any element and optimise the location of supports as well as performing cost optimisation [3]. However global optimisation is limited as the relationship between each member can become too complex for the current MOOT release.

This paper presents the development and application of a model that automates the design of reinforced concrete reservoirs using the Finite Element Method (Scia Engineer code) and a Genetic Algorithm. These are used to optimise the shape, structural element sizing and amount of reinforcement determined by least total cost using steel reinforcement and concrete volumes. The reservoir must be rectangular but may be any length and is available for many uses such as storm tanks, service reservoirs, raw water storage or an underground chamber.

The model has been called 'ResOp' (shortened from Reservoir Optimisation) and is based in Microsoft Excel due to its widespread availability and use of its VBA (Visual Basic for Applications) functionality. Some original features of ResOp are that it can either have one or two cells and columns may be included at any equal spacing and to any number required. There is also a parameter which can specify the soil stiffness at different depths of soil to suit conditions found on site. The output is a more accurate estimate of material costs (concrete and steel) which can be applied before the detailed design stage has begun. It can also be an aid at detailed design stage to find an appropriate solution efficiently without manual iteration or 'intelligent guessing'.

The model is intended to integrate a Genetic Algorithm and the latest innovations in research with the latest modelling software to make it more attractive to the wider construction industry. Currently the authors are not aware of any commercial programs that have the ability to optimise such a structure. Some less detailed programs are available but are very limited in their application.

GENETIC ALGORITHMS

Genetic Algorithms (GAs) as efficient algorithms for solution of optimisation problems have been shown to be effective at exploring large and complex search spaces in an adaptive way guided by the equivalent biological evolution mechanisms of reproduction, crossover and mutation. They are random search algorithms which have been derived based on the "Darwin's theory of survival of the fittest". A Genetic Algorithm operates on a population of trial solutions that are initially generated at random. It seeks to maximise the fitness of the population by selecting the fittest individuals from the population and using their "genetic" information in "mating" operations to create a new population of solutions. Genetic Algorithms have many advantages over the traditional optimisation methods. In particular, they do not require function derivatives and work on function evaluations alone. They have a better possibility of locating the global optimum because they search a population of points rather than a single point and they allow for consideration of design spaces consisting of a mix of continuous and discrete variables. In addition, a GA can be set in a way to provide a set of acceptable optimal or near-optimal solutions (rather than a single solution) from which the most appropriate one can be selected. The probabilistic nature of GA helps to avoid convergence to false optima [15].

Genetic Algorithm Optimisation Using Ganetxl 2006

GANetXL is an add-in for Microsoft Excel, a leading commercial spreadsheet application for Windows and MAC operating systems. Excel supports programming with Visual Basic for Applications (VBA). GANetXL is a program that uses a Genetic Algorithm to solve a wide range of single and multi-objective problems [22]. The benefit of this add-in program is its ease of use and the implementation of a GA in a spreadsheet environment that can be applied to a variety of problems.

CURRENT PRACTICE IN OPTIMAL DESIGN OF RESERVOIRS

In the past optimisation has mainly concentrated around the improvements that are made to structures by human experience and by following tables of shape ratios and selecting individually designed elements not connected to the overall structure. A popular set of tables found in 'The design of water-retaining structures' provided coefficients that could be applied to moments and forces in order to determine a generally more accurate and optimised result [2]. These ratios were based on research carried out by the Portland Cement Association of America and utilised assumptions such as the type of fixity on the walls and slabs as well as the pressure acting on the structure with the exclusion of soil conditions [4]. It suggested using these tables as a manual check to a computer technique such as FEM. The shapes of these water retaining structures were limited to rectangular, circular and conical shapes between certain size ratios.

Structurally the most efficient shapes are cylindrical and conical, this is because the wall section can be fully utilised under hoop stress from the internal liquid pressure with little bending moment. Any external pressure, so long as it is equal around the perimeter, can be efficiently supported by the concrete under compression. However treatment processes may not work effectively in a circular container which is why rectangular reservoirs are often required.

Rectangular RC reservoirs can either be jointed or monolithic in design. In both cases the optimum aspect ratio is approximately 1.5

in plan when there are two compartments (cells) for maintenance [19] and [12]. A jointed reservoir was the most popular form of construction in the past.

A jointed reservoir has movement joints to allow for thermal, flexural and tensile movement. The reinforcement usually stops either side of the joint so that a hinge is formed, which cannot transfer bending moment. Although the design can require less reinforcement (particularly in the transverse direction) these joints contain a waterbar which can be poorly constructed and which have become notorious for leakage [16]. Therefore jointed reservoirs have become less well used except in very large reservoirs because of the maintenance issues that are inherent with movement joints in contact with pressurised water.

Since the 1980s to the present, monolithic reservoirs have become more popular due to improved codes of practice that can better model crack widths, ground models can now better represent site conditions and piling has become cheaper allowing monolithic reservoirs to be built in areas previously unfeasible [19]. Steel reinforcement is continuous through the construction joint in the interface and so forces and moments can be transferred. Construction joints in a monolithic reservoir may not require any preparation before the next pour as long as the next concrete pour occurs within a relatively short timescale. If more time is required during construction then a hydrophilic strip may be placed in the centre of the wall as added security against leakage. A hydrophilic strip expands on contact with water which can seal minor breaks in the construction joint.

THE NEED FOR FURTHER OPTIMISATION

The report entitled 'Rethinking Construction' [13] noted the need to modernise by investing more in research and development of technology which was also highlighted later in 'Constructing the Team' [17]. This is occurring with BIM which instigates a need for better communication between clients, designers and contractors throughout the building lifecycle. Due to developments in technology and speed of computation, the optimisation process can be shifted closer to conceptual stage of a project. This will improve processes at the phase where changes are more likely to affect the overall project cost. By considering the structural concept early in design one is able to avoid the costs of possible redesign later in the project where changes to the design (or during construction) are more expensive. For projects such as reservoirs there is usually detailed site information early in the design or concept stage.

LATEST OPTIMISATION METHODS

There are relatively few research papers on the subject of optimising the structural elements of a building and even fewer papers on reservoirs. A two-dimensional frame was optimised by a GA and proved that it could handle discreet elements effectively [10]. The probability of crossover was 0.85 and the probability of mutation was 0.05 for a population of 50 over 50 generations. Further research was found into the optimisation of concrete structures using a heuristic flexible tolerance method, however, this research was done before mathematical optimisation algorithms were well established in civil engineering applications and does not go into detail about water tanks [21]. Sarma and Adeli [21] state that most optimisations of concrete structures were for concrete beams and girders. Conical steel water tanks have been optimised by utilising FEM and GA's particularly on elevated water towers with reductions of around 30% from standard design methods without optimisation [14]. A population size of 100 was used in the simulation. El Ansary et al. noted the superiority of Genetic Algorithms in many previous structural problems.

The optimisation of reinforced concrete reservoirs has been performed against the respective codes of practice in the country of research. Tan et al. [23] presented design of reinforced concrete cylindrical tanks using the British Standard BS8007 with a simple FEM analysis and direct optimisation techniques stating that initial feasible designs can be found using this method. This analysis was useful at the time due to the speed of this calculation, however, when compared to current FEM packages and high computational capacity this is less relevant. A more recent paper using analytical models optimised both circular and rectangular reservoirs but used simplistic optimisation methods for parametric study without the use of FEM [18]. The results showed that this was able to reduce the cost of the reservoir by shape optimisation based on the Indian codes of practice. Another relevant paper was the optimisation of a cylindrical and conical reservoir by three evolutionary algorithms and FEM. The models were based on the American Concrete Institutes building code requirements ACI 318M/318R-99. Although this optimisation uses complex algorithms, it has been limited by the fixed radius of the base and varies only by the angle of wall from vertical. Using a size of the mesh between 50 mm and 150 mm the results of this paper found the Shuffled Complex Evolution algorithm to produce the best results against the Simulated Annealing and the Genetic Algorithm. Genetic Algorithm, however, provided some similar results to the Shuffled Complex Evolution algorithm but this did take longer to run. The cost optimisation was found to be between 20% and 40% but the authors of the paper concede that there was no global reference available [1]. Recent literature has found the Shuffled Complex Evolution algorithm to be critically deficient when optimising complex nonlinear hydrological systems and improvements were made to this technique [11]. Again, no single optimisation technique has been effective for all global optimisation problems.

Optimisation found, in the limited number of research papers on the subject, was inclined towards the cost of materials and not necessarily labour, plant, formwork or temporary works [1]. This is because each of these additional costs change by region and the material costs of formwork do not necessarily represent its total cost.

Review of the current literature has highlighted a lack of application to genuine civil engineering problems of construction material costs to site constraints. Although many of the theories presented have been applied to theoretical problems they have not been used in the context of a practical design tool in real world projects of RC reservoirs. All research found did not consider partially or fully buried liquid retaining structures or 'site specific' issues. Soil conditions were not specifically considered although many were based upon elevated structures that directly connect to the structure base, which was not designed as part of the model. Soil conditions are known to be an important aspect in the design of RC reservoirs due to high loading [16] and high groundwater levels increase the risk of flotation particularly when a tank is empty [19]. Also the research found did not investigate large rectangular reservoirs with columns and so limited the length of the walls considerably.

This paper aims to utilise known site parameters to determine material volumes that can improve the accuracy of the overall project cost along with the best sizing and location of the reservoir. This is directed at a stage in the construction project where engineering concepts are usually rudimentary and cost models are not based on current site information. Site parameters such as topography and soil conditions on projects involving reservoirs are usually known reasonably early in a project to investigate concept viability (a 'yes or no' analysis). Once a site is viable then a tool such as that proposed in this paper could be utilised before detailed design. Furthermore this tool could form the beginning of an optimal concrete reservoir within detailed design, using the latest codes of practice and optimisation techniques in an easy to use interface for planning engineers and technical engineers alike.

DEVELOPMENT OF THE MODEL

The program ResOp (Reservoir Optimisation) has been created which can automatically generate a model and loadings that can be calculated using FEM analysis and optimised using a Genetic Algorithm. The program, based in MS Excel, is a spreadsheet with an input sheet containing all of the parameters and variables required to calculate the most structurally efficient rectangular reservoir. Certain variables partially dictated by the user, called chromosomes, are used in the optimisation process of the Genetic Algorithm.

Connection to Scia Engineer

Scia Engineer is a design software that uses the latest Eurocodes and the latest modelling tools which were utilised for this project. The method of transferring information into Scia Engineer is through the use of an XML document. An XML document can be compiled with any parameters or outputs from the calculated current model in Scia Engineer. Using this XML output this may then form the basis of the automated updated model.

Scia Engineer has a program that runs directly in the Windows Command Prompt executable, and therefore does not use its graphical display, which is useful for third party programs. However to view the same process that occurs with ESA_XML one can manually insert the XML document into Scia Engineer's graphical interface to prove its application. ESA_XML updates an original model, performs a specified calculation and then exports the required data into a text file (or .xls file).

An XML file is created which includes all of the variables and set engineering values shown in Fig. 1. Scia Engineer updates a basic model file to the parameters specified in the XML file and a structural solution is modelled and then calculated. An additional calculation is then performed for the design of steel reinforcing bars. Once the calculation is completed the reinforcement output is exported into a spreadsheet.

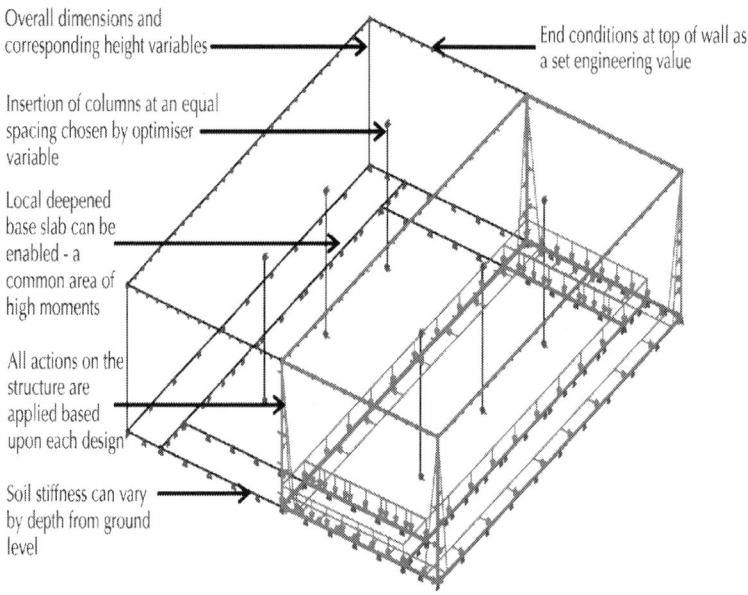

Overall dimensions and corresponding height variables

End conditions at top of wall as a set engineering value

Insertion of columns at an equal spacing chosen by optimiser variable

Local deepened base slab can be enabled - a common area of high moments

All actions on the structure are applied based upon each design

Soil stiffness can vary by depth from ground level

Figure 1: Variables and set engineering values illustration.

It is worth noting that the steel design being used is based on EC 2 part 1 and not specifically for water retaining structures ; although codes are available they were not incorporated into Scia Engineer at time of writing [6], [7], [8] and [9]. Therefore the yield strength of steel was adjusted to 200 N/mm^2 and the result combination of SLS + ULS was used. This was done to provide similar reinforcement requirements to BS8007 water retaining structures code which can be adequate as a cost model [5].

Connection with GANetXL in Excel

GANetXL used the following reservoir variables as genes in the Genetic Algorithm:

- Length of reservoir X direction,
- Length of reservoir Y direction,
- Depth of reservoir below ground level,
- Width of base slab strip surrounding external wall (for local thickening),
- Base slab edge strip depth,
- Base slab main depth,
- Column spacing (ResOp uses this value and calculates to the nearest whole column),
- External wall thickness,
- Spine wall thickness,
- Roof thickness.

The fixity condition (pinned, sliding or fixed) at the top of the walls could be included as part of the optimisation; however, this is often a preference in terms of the construction sequence. Again the requirement to divide a reservoir into two cells is usually a specification and so cannot be optimised. A storm tank or tank for a similar purpose, however, may be specified as both a single-celled or double-celled tank and so could be optimised in this way to determine the lowest cost.

These genes are all constrained to upper and lower bounds which are mostly specified by the user. It is important that the upper and lower bounds are as wide as possible so that the best solution can be sought. The single objective from the Genetic Algorithm is minimum cost.

As described earlier, the steel reinforcement output from the program is sent to a separate spreadsheet once the FEA and design have been completed. The reinforcement results are then brought into the ResOp spreadsheet (Area&Volume sheet) and the whole cost of the reservoir is calculated for the Single Objective Genetic Algorithm. A penalty is given to the chromosome for errors in the calculation and for flotation failure.

The whole process is repeated with different gene values until the certain number of generations completes. The time to process each chromosome is between 1 min and 15 min depending upon the size of the model.

Certain parameters can be changed in the generic Scia Engineer model such as the size of the mesh (which if enabled in ResOp, can reduce the length and width of the element during progression of generations) and the use of iterative calculations. These parameters will also change the length of time to calculate a single chromosome. The Scia Engineer command executable calculation is the most time intensive part of ResOp's simulation. The whole simulation program is described in Fig. 2.

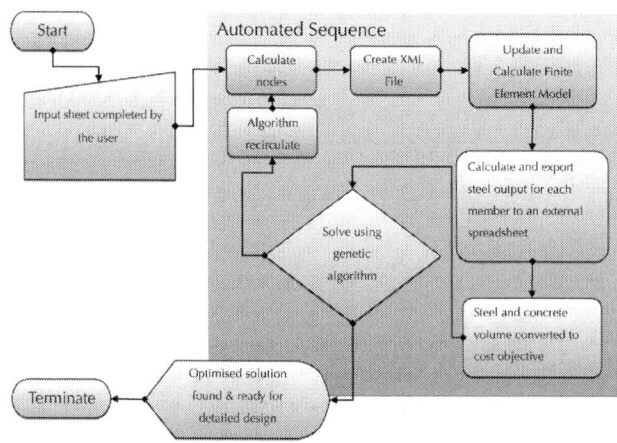

Figure 2: High-level process chart describing ResOp.

Investigation with Soil Conditions

Simple parametric modelling was performed to study the effects of soil stiffness seen in Fig. 3. The cost of the reservoir decreased with increasing soil stiffness of a reservoir with constant geometric dimensions. The volume of concrete remained the same and so only the reinforcement affected the resultant cost. These models were created to as an example of how soil conditions are important when considering cost models, particularly in poor ground conditions.

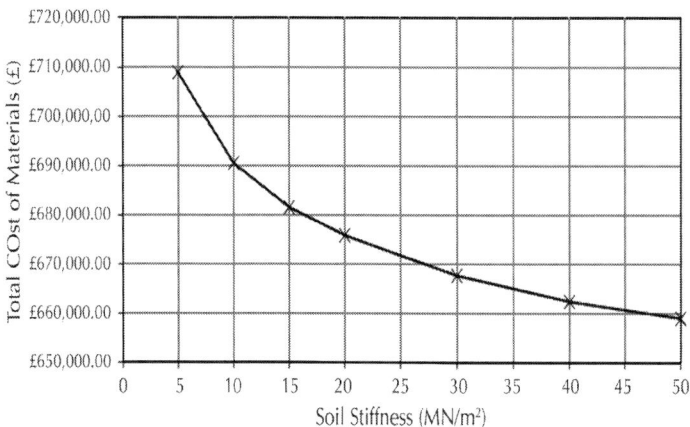

Figure 3: Effect of soil stiffness on the material cost for a 13Ml reservoir (volume: 65 m × 30 m × 7 m with spine wall).

The resulting total costs of this example decrease with increasing soil stiffness in the form of a parabolic curve. This demonstrates that the cost of materials is more influenced at lower values of soil stiffness. As an example the difference in cost between 5 MN/m² and 50 MN/m² is 7% of the total maximum cost or £49,927 as seen in Fig. 3 of a 13Ml reservoir.

Although not conclusive, this exercise suggests that the soil stiffness curve will be at an even steeper gradient at the lower stiffnesses when applying additional liquid volume without an increase in area on plan. This conclusion is suspected because an increase in load on the base and walls could extenuate the amount of reinforcing steel as the settlement of soil increases. Further analysis could be conducted using this method for both higher and lower mass using the same plan dimensions as well as using the same analysis on a reservoir without a spine wall and without a roof and fixity conditions at the top of the wall. Therefore accurate geotechnical information can make a large difference to the overall cost of a reservoir, which is why it is included as a parameter in ResOp.

MODEL SET-UP

The Genetic Algorithm was set to a population of 50, a probability of single point crossover of 0.90 and a mutation rate of 0.10 over 50 generations. These values were chosen for several reasons including previous research, from instructions for GANetXL (and experts who have used the software in other applications) and due to time constraints. A wider population and further generations would have been preferable but would have taken more time to calculate. However the results were found to be adequate for its current purpose.

Reservoir Optimisation Design Conditions for 13,000 M³ Volume

Design conditions for this reservoir are specified in and are based on a real-world example of site conditions for a water treatment works (see Table 1).

Table 1: Design conditions for the optimisation of a 13,000 m³ reservoir

Condition	Value
Storage volume	13,000 m³
Minimum head	5 m
Length 1 minimum	15 m
Length 1 maximum	85 m
Length 2 minimum	15 m
Length 2 maximum	85 m
Freeboard to top of reservoir	0.3 m
Soil stiffness between 99 and 92 m (Ground level @ 100 m)	Varies from 6 to 12 MN/m²
Base slab	Variable depth
Column fixity	Pinned (top and bottom)
Spine wall	Yes
Roof slab	Yes
Height of soil	Dictated by depth into soil
Cost of concrete (per m³)	£120.00
Cost of steel (per tonne)	£1,200.00

Average size of square mesh	1000 mm reducing to 600 mm[a]
Population size	50
Number of generations	50

Mesh convergence analysis based on cost of a similar sized reservoir.

Results of reservoir optimisation for 13,000 m³ volume

The results are compared to an existing project for a double celled reservoir 13,000 m³ in Cornwall, UK. The optimisation process, starting from a random seed, reduced the cost from £641,706 to £506,706[1]which produced a reduction of £131,308 (over 21%) as shown in Fig. 4. The graph shows that there is a general trend towards a more optimised solution with a lowering of the average cost for each generation. This proves the Genetic Algorithm is working towards a least cost solution.

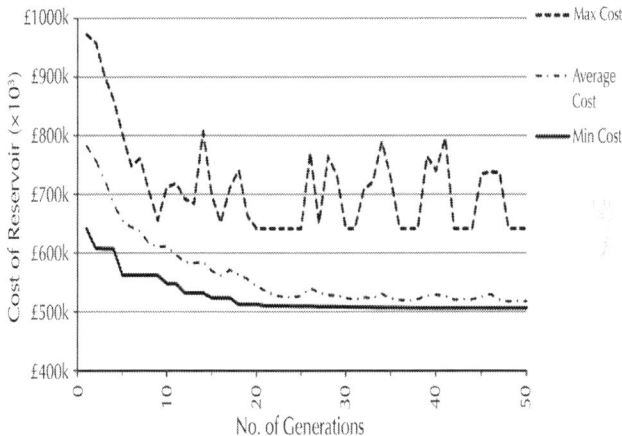

Figure 4: Graph showing a reservoir optimised for 13,000 m³ over 50 generations.

The lowest cost model is shown at every generation progressing towards an optimal solution. The last lowest cost solution remained the same for 14 generations which may indicate a global optimal solution.

Due to the length of time required to calculate a solution the size of population and number of generations was quite low. The total time used for a computer with the specification; 2.40 Quad-Core Intel i5-2430 with 6 GB RAM running Windows 7 OS, took approximately 300 h. Each chromosome took between approximately 6 min and 12 min to process. The most time-intensive process of the calculation was the output of steel reinforcement quantities which included error checking. The entire process was automated with no errors preventing the program to finish successfully.

Fig. 5 compares displacement of both the optimised and the Cornwall design models. Settlement and displacements are within tolerable levels and are generally lower than the Cornwall reservoir. Bearing stresses for both reservoirs are within tolerable levels. The optimised reservoir requires some adjustment to the soil model due to some high bearing pressures on the external walls:

$$\text{Maximum Allowable Displacement} = \frac{4500}{300}$$

$$= 15 \text{ mm (based on column spacing)}$$

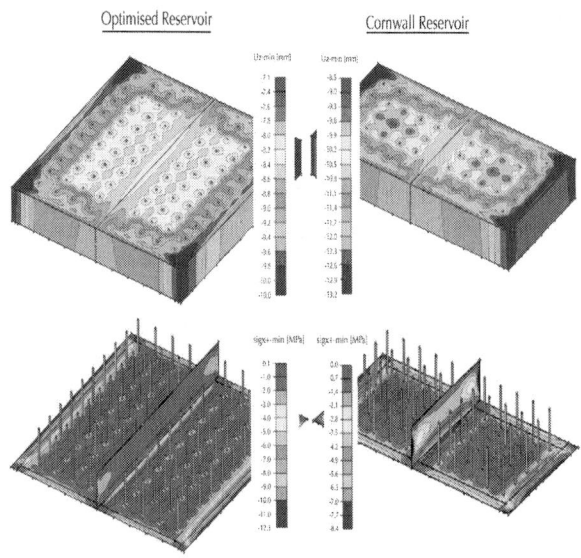

Figure 5: Vertical displacement and bearing stress for both the optimal solution and the Cornwall 13,000 m³ reservoirs.

Fig. 6 illustrates the differences in cost between the optimal reservoir design in comparison to the Cornwall reservoir model. In particular the figure shows a difference in cost of the external walls and the base slab which produces a substantial saving for the optimised reservoir. Cost of concrete is directly related to its volume and reinforcement is related to averaged values of reinforcement areas. The volume of concrete for the optimised reservoir is lower overall than the model of the Cornwall reservoir. There is also a substantial reduction in the reinforcement as the steel costs are lower for every structural element except for the roof and columns which are only marginally higher. The roof for the optimised reservoir is larger in plan area and the number of columns is greater which can account for the increase in cost.

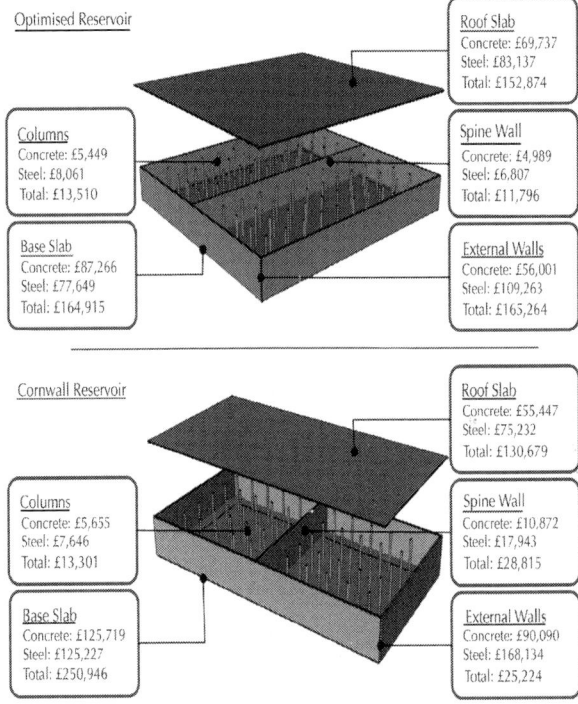

Figure 6: Cost of concrete and steel comparison by key structural elements.

When interrogating the steel reinforcement results the highest amount of steel reinforcement was found to be at the wall and the slab connections as this is where transfer of horizontal moments take place. The highest area of reinforcement was found to be 9566 mm²/m although this was very localised so the reinforcement could be B32's @ 100 mm c/c (8042 mm²/m). Much of the reinforcement requirements can be achieved with B16's @ 100 mm c/c (2010 mm²/m). This indicates that the design is buildable after further detailed checks. The reservoir in Cornwall had a combination of bars ranging from B16's @ 150 mm (1340 mm²/m) to B32's @ 150 mm (5360 mm²/m) designed using a different FEM software.

The shape and size of the reservoir optimised by ResOp (Table 2) and the one designed and built in Cornwall (Table 3) had similarities indicating engineering experience and judgement are beneficial optimisation tools. However there were differences such as the height of the optimised solution, which was 1.9 m lower, and therefore has reduced forces acting on the wall. The remaining two prominent differences are those of the plan area and the section thicknesses. Firstly the plan area of the optimised reservoir is squarer, and requires a greater plan (although wider) area than the Cornwall reservoir. Secondly the wall and slab thicknesses are greatly reduced in the optimised solution except for the roof slab, which is equal in thickness. The column spacing for the optimised design is also similar to the spacing for the Cornwall.

Table 2: Summary of output for 13,000 m³ reservoir optimised by ResOp

Storage volume	13,000 m³	
Length	46.98 m	
Width	49.48 m	
Height of reservoir	5.900 m	
Depth of base below GL	6.408 m (soil stiffness 10.6 MN/m²)	
Edge of base depth	0.40 m	
Middle base depth	0.30 m	
Number of columns	80	
Column spacing	4.70 m X direction 4.50 m Y direction	
Spine wall thickness	0.30 m	
External walls thickness	0.41 m	
Roof slab thickness	0.25 m	

Table 3: Summary of constructed service reservoir in Cornwall

Storage volume	13,000 m^3
Length	60.4 m
Width	30.6 m
Height of reservoir	7.8 m
Depth of base below GL	7.200 m approx. (soil stiffness 6–12 MN/m^2)
Edge of base depth	0.75 m
Middle base depth	0.5 m
Number of columns	50
Column spacing	5.0 m X direction; 5.1 m Y direction
Spine wall thickness	0.4 m
External walls thickness	0.6 m base 0.4 top (tapered)
Roof slab thickness	0.25 m

As shown in Table 4, 13Ml reservoir optimised using ResOp has been found to be lower in cost than the 13Ml reservoir designed and constructed in Cornwall. There is a notable difference between the model reinforcing steel quantities, which was created using ResOp without the optimisation algorithm, and the actual construction reinforcing steel quantities. This can be partly explained by the inclusion of a valve chamber, sump, upstands and staired access into both cells and other details which were not included in ResOp. Also the factor for standardising the steel reinforcement for detailing may need to be increased. Thus increasing this factor will bring the total cost of the model closer to the total actual construction cost. This is discussed in more detail later.

Table 4: Material and cost output comparison summary for 13Ml reservoirs

Double celled reservoir	Material	Quantity (approx. m^3)	Cost (approx. £)	Total cost (approx. £)
Cornwall service reservoir	Concrete	2440 m^3	£292,800	
Actual	Steel	83 m^3	£801,083	£1,000,883

Cornwall service reservoir	Concrete	2398 m³	£287,783	
Model	Steel	41 m³	£394,182	£681,965
ResOp reservoir	Concrete	1862 m³	£223,442	
Model	Steel	29.7 m³	£284,916	£508,358

Direct comparison of the two models found that substantial savings can be made using the optimised solution. However this is not the final detailed design solution and there are two points worth noting. The first is that there was a very limited design area when considering the Cornwall reservoir which would have restricted the two length genes. The ResOp solution was not restricted to such a degree because the authors felt that a more open solution would prove the programs' intelligence more so than narrowing its options. The second is that the steel reinforcement is particularly high on the wall corners horizontal steel and the centre of the walls vertical steel which could be reduced if the wall section width were increased. Construction of this nature design may be possible but perhaps not practical on site without additional engineering geometrical input. For example the engineer may make small revisions to some parameters to improve buildability.

However the results show that improvements can be applied to; the base slab and walls which could have had a reduced section thickness for similar steel reinforcement results; the walls which can be shorter; and the whole reservoir which can be more square in plan to be able to realise potential savings of over £170,000. These results have shown that real cost savings that can be made in design of reservoirs using optimisation techniques.

DISCUSSION

Size and Shape

The aspect-ratio optimised from the simulations was between 1.1 and 1.4, which was lower than the ratio found in good practice of 1.5. This figure will vary according to volume anyhow, and may not influence the design as much if the external earth pressure is insignificant in the design or other such factors.

Steel Reinforcement

The design of the steel reinforcement was carried out using Eurocode 2 Part 1 to a lower yield strength of steel, at $200 \, N/mm^2$, in order to increase the area of steel reinforcement to a value closer to liquid retaining quantities. The actual yield strength currently used in construction is $500 \, N/mm^2$. Scia Engineer did not incorporate Eurocode 2 Part 3 (Design of liquid retaining and containment structures) at the time of writing. Also two factors were included to allow for lap lengths and the application of practical steel sizes. The factor of practical steel sizes may have to be increased in future to allow for practical reinforcement detailing. There was a large cost difference between the constructed and the model reservoir. The amount of steel in the reinforcement schedule of the constructed reservoir was double that of the model. However this included stairs, parapets, sumps and a valve chamber. Also the cost of detailing steel reinforcement to satisfy good construction practice was probably more expensive than first realised. In order to lay out steel at equal centres and specify practical reinforcement the factor of steel could be increased from 1.2 to 1.5. This would mean a 50% increase in the average steel area calculated from Scia Engineer. Good detailing practice could improve this figure but more should be done to justify this factor.

Concrete Section Thickness

The section thickness of different structural elements is an important cost factor which is closely linked with reinforcement area. A thicker section can usually decrease the amount of reinforcement required in the section although the dead weight will increase. The section thicknesses observed were usually thinner than those used in practice, and this can increase the reinforcement area and may make such solutions impractical for construction. However ResOp can limit such solutions to a certain extent by restricting the genes which specify section thicknesses.

Intelligent Design

Although a certain section can be quite thin over the majority of the wall, there are some very localised areas (usually at the connections to other structural elements) that may have impractical steel area requirements. ResOp has little intelligence with regard to these problems. The steel reinforcement requirements found from ResOp are currently averaged over a whole structural element (such as the North wall or roof). In order to make these results more realistic the results may need to be skewed more to the side of higher reinforcement to request greater section thicknesses. However further trials will have to be done to determine the amount of skew.

Depth into Soil

It was observed that the depth into soil was 500 mm deeper than the total depth of the reservoir. This is possibly due to the increased soil stiffness found at this depth. This depth into soil has been taken into account for the external walls but has not increased the load directly onto the roof. This has been thought to insubstantially increase the total cost. In practice this may be a design requirement of the site and can be considered further at detailed design.

CONCLUSIONS

Economies to a rectangular reinforced reservoir have been found through a Genetic Algorithm based on a combination of commercially available software. The cost savings could vastly outweigh the cost of the program and its components, particularly as most are readily available to many companies. The program ResOp does not eliminate the need for a structural engineer but can be used as a tool to contribute to the design process, particularly at an earlier stage in design. Some solutions produced may not be viable in terms of cost or buildability, therefore an experienced structural engineer will be required at detailed design stage or earlier. The process to use ResOp requires both project managers and civil and structural engineers, as before, but provides them with a deeper understanding of construction costs when considering a design. To produce a least cost solution at the touch of a button is no longer a future technology and this should be harnessed more in civil engineering, not just as mechanical and electrical, aerospace and naval engineering. The construction industry is undergoing an advancement with the use of BIM that should make tools used by ResOp more available due improvements in software and better site investigations and topographical surveys.

Currently the length of time to calculate a solution using ResOp is excessive but this will improve with more efficient coding and using more powerful computers. Also the population and number of generations are low for this type of analysis and should be increased with more efficient finite element analysis and algorithm software. Further resources could create parallel solutions and may reach an optimised solution faster.

The emerging technologies used in ResOp are increasing in popularity and the tool was programmed using well founded computer languages. Design in structural engineering should combine both human and computer intelligence; and replicating previous designs without considering site parameters should no longer be practiced. A deeper understanding of cost during preliminary design and detailed design will enable savings to be made throughout the design and construction phases of a project.

ACKNOWLEDGMENTS

The authors would like to thank Pell Frischmann and CADS for information and support in the creation of this paper.

REFERENCES

1. Barakat SA, Altoubat S. Application of evolutionary global optimization techniques in the design of RC water tanks. Eng Struct 2009; 31:332–44.

2. Batty I, Westbrook R. The design of water-retaining structures. J. Wiley & Sons; 1991.

3. Blazek R, Novak M, Roun P. Scia engineer MOOT: automatic optimisation of civil engineering structures. Nemetschek; 2011.

4.   British Cement Association. Guidance on economic design, detailing and specification. UK: British Cement Association; 1994.

5.   British Standards Institute. BS8007 – design of concrete structures for retaining aqueous liquids. UK: BSI; 1987.

6.   British Standards Institute. Eurocode – basis of structural design. UK: BSI; 2002.

7.   British Standards Institute. Eurocode 2: design of concrete structures. In: Part 1-1: General rules and rules for buildings. UK: BSI; 2004.

8.   British Standards Institute. Eurocode 2 – design of concrete structures. In: Part 3: Liquid retaining and containment structures. UK: BSI; 2006a.

9.   British Standards Institute. UK National Annex to Eurocode 2: design of concrete structures. In: Part 3: Liquid retaining and containment structures. UK: BSI; 2006b.

10.   Camp C, Pezeshk S, Cao GZ. Optimized design of two-dimensional structures using a genetic algorithm. J Struct Eng – ASCE 1998; 124:551–9.

11.   Chu W, Gao X, Sorooshian S. Improving the shuffled complex evolution scheme for optimization of complex nonlinear hydrological systems: application to the calibration of the Sacramento soil-moisture accounting model. Water Resour Res 2010:46.

12.   Crocker THW. Design and maintenance of service reservoirs, vol. 46. Exeter: South West Water, UK; 2008.

13.   Egan J. Rethinking construction – the report of the construction task force. Proc Inst Civil Eng-Munic Eng 1998; 127: 199–203.

14.   El Ansary AM, El Damatty AA, Nassef AO. A coupled finite element genetic algorithm technique for optimum design of steel conical tanks. Thin-Walled Struct 2010;48: 260–73.

15.   Holland JH. Genetic algorithms. Sci Am 1992; 267:66–72.

16.   Jones GM, Randy Nixon RA, Parks RR, Sanks RL, Blanchard CT, Bouthillier PH, et al. Summary of design considerations. In: Garr MJ, Pe LS, Robert PhD, George T, Bayard PE, Bosserman li E, editors. Pumping station design. Burlington: Butterworth-Heinemann; 2008. p. 25.1–25.52 chapter 25..

17. Latham M. Constructing the team: joint review of procurement and contractual arrangements in the United Kingdom Construction Industry: final report. H.M. Stationery Office; 1994.

18. Mohammed HJ. Economical design of water concrete tanks. Eur J Sci Res 2011; 49:10.

19. Ratnayaka DD, Brandt MJ, Johnson M. Treated water storage. In: Twort's, editor. Water supply, vol. 744. Butterworth-Heinemann; 2009.

20. Roberts D. Autogenous healing: the self-sealing of fine cracks. Concrete Society; 2003.

21. Sarma KC, Adeli H. Cost optimization of concrete structures. J Struct Eng – ASCE 1998;124: 570–8.

22. Savic´ DA, Bicik J, Morley MS. A DSS generator for multiobjective optimisation of spreadsheet-based models. Environ Modell Softw 2011; 26: 551–61.

23. Tan GH, Thevendran V, Das Gupta NC, Thambiratnam DP. Design of reinforced concrete cylindrical water tanks for minimum material cost. Comput Struct 1993; 48: 803–10.

Natural Gas Treating by Selective Adsorption: Material Science and Chemical Engineering Interplay

Marco Tagliabue[a], David Farrusseng[b], Susana Valencia[c], Sonia Aguado[b], Ugo Ravon[b] Caterina Rizzo[a], Avelino Corma[c], and Claude Mirodatos[a]

[a]Eni S.p.A, Refining & Marketing Division, via Felice Maritano 26, 20097, San Donato Milanese, Italy

[b]Université Lyon 1, IRCELYON, Institut de Recherches sur la Catalyse et l'Environnement de Lyon, UMR CNRS 5256, avenue Albert Einstein 2, 69626 Villeurbanne, France

[c]Instituto de Tecnología Química, UPV-CSIC, Universidad Politécnica de Valencia, Avenida de los Naranjos s/n, 46022, Valencia, Spain

ABSTRACT

The paper addresses current needs in Natural Gas (NG) treating. Basic principles of Pressure Swing Adsorption (PSA) separation processes are described. A state of the art of microporous adsorbents in the frame of NG treating is given. It includes reference and advanced zeolites, carbon based materials and Metal-Organic Frameworks (MOFs). The pros and cons of each material category are discussed. Guidelines to develop on-purpose materials are given from thermodynamics and material state of the art. Finally, PSA applicability to inert (nitrogen and carbon dioxide) rejection from NG is discussed.

SCOPE OF THE REVIEW

Gas treating technologies can be roughly divided into two main categories [1] and [2]: (i) *separation*, with contaminant concentration of about 10 wt.% or higher in feed; (ii) *purification*, with contaminant concentration less than about 3 wt.% in feed. Pressure Swing Adsorption (PSA) represents a state of the art technology both in separation and purification of gaseous mixtures [3], [4], [5], [6], [7], [8] and [9]. As in all adsorption technologies, the nature of feed and products drives the choice of the most adequate adsorbing material and process design. For this reason, a strong integration between process engineering and material science is requested in PSA development and optimisation.

The paper addresses current needs in *Natural Gas* (NG) treating. Basic principles of adsorption processes including adsorbent–adsorbate interactions and current PSA technologies are described. A state of the art of microporous adsorbents in the frame of NG treating is given. It includes reference and advanced*zeolites, carbon-based materials* and *Metal-Organic Frameworks* (MOFs). Finally, PSA applicability to inert (nitrogen and carbon dioxide) rejection from NG is discussed.

NATURAL GAS: FROM THE RESERVOIR TO THE MARKETPLACE

Resources and Specifications

The progress of the international energy demand shows a 1.7% average annual growth in the 2005–2020 period. This growth concerns all energy sources, although fossil fuels will still rule the energy scene for the next fifteen years.

NG demand will account for the highest growth rate and in 2020 it will exceed that of coal, that will be penalised by the increasing restrictions in pollutant emissions (especially in Europe).

Novel transport technologies, the remarkable reserves found, the lower overall costs and the environmental sustainability all point to NG (less polluting than oil and coal and now used in more efficient plants) as the primary energy source in the near future [10] and [11].

Reservoirs are frequently far from final markets (Table 1): as a consequence, NG has to be transported either by pipelines as gaseous mixture containing at least 75 vol.% of methane or by tankers as *Liquified Natural Gas* (LNG), containing at least 85 vol.% of methane [12]. The choice between the two transportation technologies depends mainly on the distance and on the volume of gas to be transported. According to an Eni on-shore case-study, LNG could be considered the preferred choice in case of relatively small fields (less than 1 Tcf $y^{-1} = 2.83 \times 10^{11}$ m^3 y^{-1}) and long distance (more than 3000 miles = 4800 km). NG is frequently identified according with its origin and/or chemical composition. Classification by origin provides three main categories: (i) *non-associated gas* which is not in contact with oil; (ii) *gas-cap associated gas* overlying the oil phase in the reservoir; (iii) *associated gas* dissolved in the oil at the reservoir conditions (*solution gas*).

Table 1: 2007 world gas reserves (referred to world gas reserves = 1.82×10^{14} m^3) compared to 2005 world gas production (referred to world gas production = 2.84×10^{12} m^3) and 2005 world gas consumption (referred to world consumption = 2.82×10^{12} m^3) [11]

Area	Reserves [%]	Production [%]	Consumption [%]
Western Europe	3.0	10.4	17.6
Industrialised Asia and Pacific	1.6	1.8	6.0
North America	4.2	24.4	24.7
Central Europe	0.3	0.8	2.7
Eastern Europe	27.0	22.4	18.5

Central Asia	4.6	5.3	3.6
Middle East	40.3	10.3	8.9
Africa	7.9	6.5	3.0
Developing Asia and Pacific	7.2	11.7	8.9
Latin America	4.0	6.4	6.2

As produced from gas fields, NG generally contains variable amounts of several contaminants such as water, light paraffins, aromatics, carbon dioxide, nitrogen and sulphur compounds. Minor amounts of helium (less than 1 vol.%) and mercury (generally 5–300 μg Nm^{-3}, in few cases more than 1000 μg Nm^{-3}) can be also present. Furthermore, enhanced hydrocarbon recovery operations by nitrogen and/or carbon dioxide injection into reservoirs modify the chemical composition of produced gas. NG can be also classified into *dry* or *wet* depending on the amount of C_{2+} hydrocarbons (wet gas for C_{2+} content higher than 10 vol.%) and into *sweet* or *sour* depending on the amount of acid gas (sour gas for hydrogen sulphide content higher than 1 vol.% and/or carbon dioxide content higher than 2 vol.%). Chemical composition determines the operations requested to meet specifications requested for NG transportation and final processing (Table 2).

Table 2: Typical compositional specifications on feed to LNG plant and on pipeline gas (Total Sulfur refers to H_2S + carbonyl sulphide, COS + organic sulphur) [12] and [13]

Impurity	Feed to LNG Plant	Pipeline Gas
H2O	<0.1 ppmv	150 ppmv
H2S	<4 ppmv	5.7–22.9 mg Sm−3
CO2	<50 ppmv	3–4 vol.%
Total Sulfur	<20 ppmv	115–419 mg·Sm−3
N2	<1 vol.%	3 vol.%
Hg	<0.01 mg/Nm3	–
C4	<2 vol.%	–
C5+	<0.1 vol.%	–
Aromatics	<2 ppmv	–

Unconventional hydrocarbon sources such as *CoalBed Methane* (CBM) and *LandFill Gas* (LFG) have recently drawn energy companies' attention. The gaseous mixtures entrapped in coal beds are mainly composed by methane (up to 80–99 vol.%) and minor amounts of carbon dioxide, nitrogen, hydrogen sulphide and sulphur dioxide [14]. The gas produced by waste decomposition in landfills is a complex gaseous mixture containing carbon dioxide as the main contaminant [15]. In general, before supplying to processing plants and downstream customers *gas conditioning* operations are requested [12], [13], [16],[17] and [18]. Water content has to be reduced to levels that prevent corrosion, hydrate formation and freezing in cryogenic equipments. Indeed, the formation of hydrates from water and hydrocarbons is considered the primary cause of plugging of transmission lines. Until today, the most applied water separation technology remains scrubbing with TriEthyleneGlycol (TEG), followed by scavenging on solid adsorbents (zeolite, silica, silica–alumina, alumina pure or mixed together, activated carbons) [19]. After dehydration, TEG is regenerated at high temperature (about 473 K) and recycled to scrubbers. Hydrocarbons heavier than methane contribute to NG heating value. On the other hand, C_{3+} hydrocarbons can cause problems: either pressure or temperature variations can cause their fall out of the gas phase. C_{3+} hydrocarbons can also cause plugging of downstream valves and pipes or fouling of other equipments (*i.e.* gas separation membranes). Their content is generally adjusted by gas chilling and/or scrubbing with liquid hydrocarbons and successive scavenging on the same solid adsorbents used for dehydration.

The removal of inert gases (mainly nitrogen but also helium) and acid gases (such as carbon dioxide and hydrogen sulphide) is of considerable importance inasmuch as they can be present to a significant extent. NG containing hydrogen sulphide in proportions higher than 10 vol.% are not very common and many gases contain practically no hydrogen sulphide. Conversely, nitrogen and carbon dioxide are common NG contaminants, with average proportions in the range 0.5–5.0 vol.% for nitrogen (with peaks of over 25 vol.%) and 0.5–10 vol.% for carbon dioxide (with peaks up to 70 vol.%). When sulphur compounds and carbon dioxide levels are too high, they need to be reduced in order to avoid formation of solids in cryogenic units and steel pipes corrosion. Furthermore, both nitrogen and carbon dioxide can be considered *inert gases* with no heating value: for this reason, they must be removed to low levels (Table 2) before distribution to final users. Nitrogen and helium rejection from NG are usually operated by cryogenic fractional distillation. Cryogenic *Nitrogen Rejection Units* (NRUs) are considered economically acceptable for gas flows exceeding 15 MMscf d^{-1} = 4.25 × 10^5 Sm^3 d^{-1} [20], although higher flow ranges are recommended (50 MMscf d^{-1} = 1.42 × 10^6 Sm^3 d^{-1}) [18]. Acid gas bulk removal (*gas sweetening*) is mainly performed either by aqueous amine or organic solvent scrubbing. The acid gas saturated (*rich*) fluid is regenerated by high temperature stripping (less than about 400 K, in case of amine) or by pressure reduction (in case of solvents) and then recycled to scrubbers. These treatments have to be run carefully, in order to reduce risks deriving from dangerous chemical handling and toxic waste production.

Opportunities from PSA Technologies for Natural Gas Treating

So far, the exploitation of highly contaminated gas streams (*poor gases* or *low-Btu gases*) has not been considered economically attractive because of high capital and operative costs of current gas treating technologies. As a consequence, these gaseous mixtures are frequently not converted into energy [21]. In this context, among the emerging gas treating technologies, PSA processes have drawn market attention because of their intrinsic eco-compatibility and flexibility:

- PSA technology is based on the use of regenerable solid adsorbents. According to that, it does not require the management of chemicals (e.g. amine, solvents), thus providing considerable environmental benefits;
- adsorbent regeneneration does not require heating. Consequently, PSA process energy intensity is low;
- PSA units can be easily downsized to skid-mounted modules suitable for the exploitation of small gas reservoirs.

PSA GENERAL FEATURES

PSA basic concepts are described in two patents granted in the late 1950s [22] and [23] and since then, R&D effort in this field has been continuous [4] and [7]. Sizes of commercial PSA units range from small devices (about 300 scf d^{-1} = 8.50 Sm^3 d^{-1}) for the production of 90 vol.% oxygen from air for medical use, to large refinery plants (about 100 MMscf d^{-1} = 2.83 × 10^6 Sm^3 d^{-1}) for the production of high purity hydrogen from *Steam Methane Reforming* (SMR).

Like all adsorption separation processes, PSA involve two basic steps:

- during *adsorption step*, certain components of a gaseous mixture are selectively adsorbed on a porous solid. This operation, performed at relatively high pressure by contacting the gaseous mixture with the adsorbent in a packed column, produces a gas stream enriched in the less strongly adsorbed component of the feed mixture (the *raffinate*). After a given time of operation, the adsorbing bed approach saturation and regeneration is requested;

- during *regeneration* or *desorption* step the adsorbed components are released from the solid by lowering their gas phase partial pressures inside the column. After this operation, the adsorbent is ready to be employed in a further cycle. The gaseous mixture obtained from regeneration (the *extract*) is enriched in the more strongly adsorbed components of the feed.

In practice, several columns are operated in a swing-mode to make the process continuous and additional steps are added to the basic cycle, in order to maximise productivity and energy saving.

It is common practice to include a low pressure purge step in the cycle in order to promote adsorbent regeneration. Purge is generally operated by recycling part of raffinate (*purgegas* or *sweep gas*). According to that, raffinate is obtained at about the same pressure of feed, in a pure form while the extract is discharged as secondary product, in impure form and at pressure lower than feed. The whole cycle last minutes or even seconds and is operated under approximately isothermal conditions. The *working capacity* is the difference in loading between the points corresponding to adsorption and regeneration pressures on the same adsorption isotherm. It has to be noticed that in cases of strongly adsorbed species heating at high temperature (*e.g.* around 573 K for release of water from zeolites in NG dehydration processes) is the only regeneration option. This strategy is referred as *Temperature Swing Adsorption*(TSA).

NG is often available at wellhead at high pressure. At least in principle, by using adsorbents able to capture contaminants, PSA processes can produce pure methane at high pressure as raffinate, thus reducing further compression work before transmission to downstream customers.

ADSORPTION FUNDAMENTALS

The essential requirement of adsorption separation processes is an adsorbent that preferentially adsorbs a family of related components from a mixed feed. Adsorbent selectivity may depend on difference in adsorption at equilibrium (*equilibrium selectivity* or *thermodynamic selectivity*) or on a difference in adsorption rates (*kinetic selectivity*). Kinetic selectivity is possible when a great difference among adsorption/desorption rates of different components exists. This is due to sterically hindered diffusion through pores characterised by pore mouth diameter comparable with molecular size of fed species. Difference in rates may be so great that the slower diffusing species are almost excluded from the adsorbent (*size-selective sieving*). Currently, nitrogen separation from air by using small pore zeolites or carbon molecular sieves is the only commercial PSA process based on kinetic selectivity. Therefore, both solid–fluid phase interactions and diffusion through adsorbent pores must be considered during material selection and tailoring [3], [6], [24] and [25].

Gas–Solid Interactions

Adsorption (*physisorption*) is based on attraction forces among the solid phase and the species constituting the gas phase, with relatively low adsorption heat. Usually, $qads < 2qvap$, where $qads$ and $qvap$ represent, respectively, adsorbate heat of adsorption and vaporisation. Adsorption forces can be categorised into two main groups: (i) *van der Waals forces*, directly correlated with adsorbate molecular polarisability and (ii) *electrostatic forces* such as polarisation forces, surface field–molecular dipole interactions and surface field gradient–molecular quadrupole interactions. Adsorption is usually promoted by synergies among these different kinds of interactions. Key physico-chemical properties of common NG components are reported in Table 3. Carbon dioxide and nitrogen adsorption on polar surfaces (e.g.zeolites) is mainly promoted by surface field gradient–molecular quadrupole interactions. Conversely, adsorption of large non-polar molecules (e.g. hydrocarbons) is essentially due to their molecular polarisability.

Table 3: Physico-chemical properties of some molecular species present in NG (hydrogen and carbon monoxide are reported for comparison): : *kinetic diameter*, : *polarisability, μ: dipole moment*, : *quadrupole moment*, T_c: *critical temperature*. Quadrupole moment indicative values have been computed by *Density Functional Theory* (DFT) method

Molecule	[Å]	[Å3]	μ [D]	[D·Å]	T c [K]
CH4	3.80	2.448	0.000	0.02	190
N2	3.64	1.710	0.000	1.54	126
CO2	3.30	2.507	0.000	4.30	304
H2O	2.65	1.501	1.850	2.30	647
H2S	3.60	3.630	0.970	3.74	373
He	2.60	0.208	0.000	0.00	5
H2	2.89	0.787	0.000	0.43	39
CO	3.76	1.953	0.112	2.04	133
Reference	[26]	[27]	[27]	[27]	[28]

Adsorbents able to promote electrostatic interactions (as well as hydrogen bonding) are generally referred as *hydrophilic*. In general, they are able to adsorb small polar molecules (like water) much more strongly than would be expected simply from van der Waals forces alone. Common hydrophilic adsorbents are zeolites, silica and alumina. Conversely, adsorbents operating exclusively by van der Waals forces (e.g.most activated carbons) are referred as *hydrophobic*. According with the gas–solid interactions described above, thermodynamic selectivity can be optimised by tailoring adsorbent physico-chemical characteristics. Zeolite polarity can be tuned at the synthesis step (e.g. by choosing the Si/Al molar ratio and counter cations) or by post-synthesis treatments [29]. Besides, polar groups such as carboxylic species can be introduced on activated carbons by oxidation treatments [30].

Adsorbent Porous Texture

Adsorbent–adsorbate interactions play an important role for the very first adsorption layers. However, adsorption mechanism and hence separation performances are essentially determined by the porous texture of adsorbent. IUPAC classification categorises pores into micro-, meso- and macropores according to their *pore size* (*d, i.e.* diameters of cylindrical pores or distance between the sides of slit-shaped pores). This classification corresponds to different adsorption mechanisms, although the ratio adsorbate molecule size *vs.* pore size is the effective discriminating factor. Adsorption in *micropores* ($d \leq 20$ Å) takes place by *micropore filling* [31] and [32]. In this case, adsorbent–adsorbate interactions are greatly enhanced since gas molecules and pore sizes are comparable and each gas molecule experience the force field generated by pore walls. In addition, molecular sieving can take place if one component is larger than pore opening. Microporous zeolites and carbons commonly used adsorbents as described in Sections 4.1 and 4.2.

Regarding adsorption both in *mesopores* (20 Å $< d \leq 500$ Å) and in *macropores* ($d > 500$ Å), adsorbate molecules are organised by multiple layers according to the proposed mechanistic models (*e.g.* the Brunauer–Emmett–Teller one [33]). Molecules belonging to the layer contacting the solid surface are strongly attracted while molecules in the central region of the pore are essentially free from the force field. Adsorbate partial pressure inside mesopores is higher than outside. As partial pressure exceeds critical value (if adsorption temperature is below the critical value), bulk adsorbate condensation takes place inside the pore. This phenomenon is referred as *capillary condensation*.

Together with surface interactions, capillary condensation causes water and C_{3+} hydrocarbon adsorption on mesoporous absorbents. For this reason, mesoporous silica, alumina and silica–alumina are extensively used in NG dehydration and C_{3+} hydrocarbon removal (Section 1). Besides, M41S materials have been recently proposed for gas treating [34].

Macropores ($d > 500$ Å) behave in gas adsorption as open surfaces. Thus, their contribution to adsorption capacity is generally negligible and their main role is to facilitate transport within adsorbent particles. According to that, most commercial adsorbents consist of aggregated microporous grains, usually with a binder, to form macroporous hyper-structures. The binder also provides mechanical resistance (e.g. to pressure shocks) to adsorbent particles.

Adsorption Equilibrium and Heat of Adsorption

The concept of adsorption equilibrium is deeply involved in the evaluation of adsorbent specific capacity, selectivity and regenerability (working capacity). Equilibrium *adsorption isotherms* $n_i = f(p_i, T)$ (where n_i is the amount of component i adsorbed at temperature T and at partial pressure p_i) and *heat of adsorption* represent essential input data for PSA process modelling [35].

Heat of adsorption is a measure of the strength of interactions between adsorbate and adsorbent (adsorption is an exothermal phenomenon). For adsorbents characterised by energetically heterogeneous surfaces (e.g. most zeolites) heat of adsorption is higher at low loadings, describing interactions on strongest sites. Thus, it is properly referred as *isosteric* (*i.e.* at a definite loading) heat of adsorption, q_{st}. From process engineering point of view, heat of adsorption is a measure of the energy required for adsorbent regeneration and provides an indication of temperature variations that can be expected on the bed during adsorption (and desorption) under adiabatic conditions. Table 4 and Table 5 provide representative heat of adsorption values for some NG components referred to common adsorbents. Due to high adsorption heat, variations of about 30 °C are reported on zeolitic materials for carbon dioxide capture from NG [36]. These temperature variations are detrimental for PSA efficiency since temperature rising during adsorption step decreases adsorbate uptake and bed temperature decrease during desorption step reduces adsorbate release.

Table 4: Isosteric heat of adsorption extrapolated at zero coverage (q_{st}) of some NG components on commercial activated carbons and zeolites at about 303 K

Adsorbent	q_{st} [kJ mol−1]			Reference
	CH4	CO2	N2	
Norit Extra	20.60	22.00	–	[37]
Calgon BPL	16.10	25.70	–	[37]

Kensai Maxsorb	16.60	16.20	–	[37]
A'dall A10	16.20	21.60	–	[37]
Osaka Gas A	18.30	17.80	–	[37]
Silicalite	20.90	27.20	17.60	[38]
NaZSM-5	26.50	50.00	24.10	[39]
NaX	19.20	49.00	19.90	[39]

Table 5: Average heat of adsorption (q) of some NG components on various adsorbents at about 303 K

Adsorbent	q [kJ mol−1]					Reference
	CH4	CO2	N2	H2O	C2H6	
NaX	13.81	34.31	16.74	51.46	31.38	[40]
KClinoptilolite	25.10	39.75	25.10	–	29.29	[40]
CaClinoptilolite	15.06	25.10	20.08	–	10.46	[40]
MgClinoptilolite	29.98	–	19.99	–	–	[41]
SrETS-4	14.67	–	21.20	–	–	[41]
-Al2O3	10.46	33.47	8.37	48.53	17.57	[40]

Selectivity

Equilibrium selectivity is the key parameter to evaluate adsorbent separation ability. It is based on differences in affinities of the adsorbent for the different species constituting the fluid phase. Given an adsorbent and a gaseous mixture in which N is the more strongly adsorbed component and M the predominant but less strongly adsorbed one, equilibrium selectivity is generally expressed by using the *separation factor* $_{N,M}$:

$$\alpha_{N,M} = \frac{n_N \, p_M}{n_M \, p_N} \approx \frac{K_N}{K_M}$$

(1)

where n_N, p_N, n_M, p_M are values obtained from pure component isotherms. It may be shown that in many cases $\alpha_{N,M}$ can be computed as the ratio among N and M Henry's constants (K_i) thus often referred as *Henry's selectivity*. For PSA applications, $_{N,M}$ values comprised between 2 and 104 may be considered acceptable [9]. For the same adsorbent, adsorbed and desorbed amounts can differ significantly from different conditions of pressures and temperatures. Furthermore, kinetic effects have to be considered when relevant. According to that, additional selectivity criteria have been suggested [42], [43] and [44].

Selectivity *vs.* Regenerability

An energy intensive regeneration step is the price you have to pay in case of very selective adsorbents. For this reason, adsorbent design main objective should be to find out an optimal trade-off between selectivity and regenerability. Table 6 contains a collection of separation factors referred to different couples of gases on Calgon BPL™ activated carbon and CaNaA zeolite (commercial 5 A molecular sieve) [8].

Table 6: Separation factors N M on Calgon BPL™ activated carbon and Ca-NaA zeolite (commercial 5 A molecular sieve) at 303 K [8]

Mixture N–M	N,M	
	BPL™ Activated Carbon	CaNaA Zeolite
CO2–CH4	2.5	195.6
CO2–CO	7.5	59.1
CO2–N2	11.1	330.7
CO2–H2	90.8	7400.0
CO–CH4	0.33	3.3
CO–N2	1.48	5.6
CO–H2	12.11	125.0
CH4–N2	4.5	1.7
CH4–H2	36.6	37.8
N2–H2	8.2	22.3

Zeolite CaNaA is much more selective than activated carbon in carbon dioxide separation from other gases. On the other hand, the great affinity of CaNaA zeolite for carbon dioxide make extremely difficult carbon dioxide displacement by sweeping with other gases (e.g. methane, hydrogen), as requested in typical PSA processes (Section

2). Effective carbon dioxide desorption from CaNaA zeolite can only be obtained by applying more complex or energy intensive strategies like vacuum or high temperature heating (TSA). Regarding nitrogen rejection from NG, it is evident that both Calgon BPL™ activated carbon and CaNaA zeolite show poor selectivities. Although this is merely a case study, it depicts a general behaviour: both methane and nitrogen are able to interact weakly with most adsorbents. Furthermore, their molecules show similar kinetic diameters (σ_{N2} = 3.64 Å, σ_{CH4} = 3.80 Å), making molecular sieving extremely difficult. Thus, very few adsorbents are able to selectively adsorb nitrogen from NG (Section 4.1).

ADSORBENT DESIGN AND STATE OF THE ART

Adsorbent development is an essential step in PSA process design, as outlined in previous Sections. Adsorbents suitable to PSA processes have to satisfy several requisites: (i) selectivity; (ii) regenerability by pressure reduction; (iii) specific capacity; (iv) fast interparticle diffusion; (v) chemical and physical stability; (vi) low cost per unit volume; (vii) reasonable packing density to avoid oversized vessels. It is important to stress that adsorbent cost is a key parameter to be considered for industrial application. Common A, X and Y zeolites are extensively employed in gas industry, although the use of more sophisticated structures is frequently proposed in literature as a way to improve separation performances. Zeolite and carbon adsorbents have been successfully used for inert rejection from raw NG streams whereas MOFs are emerging adsorbents that are gathering great attention because of their outstanding pore volumes. These three categories of adsorbents are described in Sections 4.1, 4.2 and 4.3.

Microporous Zeolites and Aluminophosphates

Zeolites found in nature and synthesised ones are obtained under hydrothermal conditions and can be divided in those having low Si/Al molar ratio (from 1 to 5) and those of high Si/Al molar ratio (from 5 to ∞), being the formers much more hydrophilic materials [26]. Zeolites are tridimensional aluminosilicate frameworks constituted by Si and Al tetrahedra linked through bridging oxygen atoms giving rise to a regular distribution of pores and cavities of molecular dimensions (Fig. 1). In low Si/Al molar ratio zeolites, the negative charge generated by aluminium in the framework is compensated by small cations such as alkali and alkaline-earth metal cations, whereas the use of large organic molecules allows obtaining higher Si/Al molar ratio zeolites. By means of a great variety of organic molecules based on quaternary ammonium cations as *Structure Directing Agents* (SDAs), several existing zeolites have been obtained with higher Si/Al molar ratios and interestingly, a large number of new zeolitic materials have been synthesised. Up to now, 191 different zeolitic materials with known structure exist [45] and this number is even larger since zeolites in which the structure has not been solved are reported in the open and patent literature. Aluminophosphates (AlPOs) are related materials with similar structures than zeolites constituted by Al and P tetrahedra.

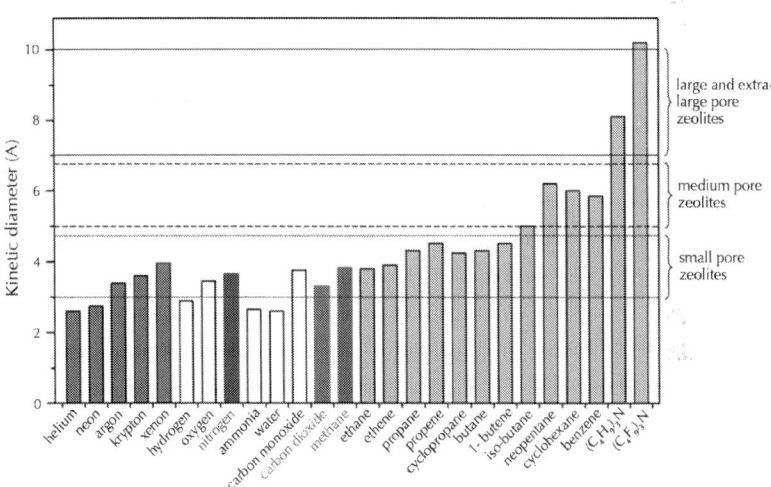

Figure 1: Kinetic diameter of several molecules (taken from ref. [26]) compared with the pore diameter of different types of zeolites.

Discovery of zeolitic microporous adsorbents has represented one of the major breakthroughs in the field of adsorption and separation of gas mixtures. The great availability of zeolite structures with different pore size and topology, together with the possibility of preparing them in a wide range of chemical compositions make zeolites very useful adsorbents for gas separations, getting chances for the selection of the most appropriated material for a given separation process. As a consequence of their well defined crystalline structures, these materials have uniform pore sizes in the range of molecular dimensions (3–10 Å) achieving microporosity volumes up to 0.35 cm^3 g^{-1}. Furthermore, the benefit of the high thermal, hydrothermal and chemical stability, allows their use as adsorbents and catalysts in many processes[46] and [47].

Zeolites can be classified according to different criteria. The most commonly used is the one referring to the*dimensions of the pore apertures*. According to that, zeolites can be divided in the following groups:

- *small pore zeolites*: constituted by channels delimited by 8 *Member Rings* (8 MR) with pore diameters around 4 Å. In this group, zeolitic materials such as zeolite A (LTA) and chabasite (CHA) are the most common;

- *medium pore zeolites*: constituted by channels delimited by 10 MR and pore apertures around 5–6 Å. Zeolites ZSM-5 (MFI) and ferrierite (FER) belong to this group;

- *large pore zeolites*: constituted by channels delimited by 12 MR with pore diameters around 7 Å. Faujasite (FAU) type of zeotites (X and Y) and beta (BEA) are large pore ones;

- *extra-large pore zeolites*: constituted by channels delimited by rings of more than 12 MR and pore apertures larger than 7 Å. These zeolitic materials have been obtained in the latest years and examples of them are CIT-5 (CFI) and ITQ-33 zeolites.

The *shape of the pores* is another important factor, since there are zeolites with pore apertures delimited by the same number of tetrahedra but with different shapes, making them to behave quite differently when adsorbing molecules. For instance, zeolite A (LTA) and analcime (ANA) are both small pore zeolites, being the former comprised by circular pores of 4.1 Å and the latter consisting of elliptical pores with a smaller dimension of 1.6 Å that makes it useless for adsorption of molecules.

Other criterion for classifying zeolitic materials is the *dimensionality of their channels*. Thus, zeolite structures comprised by one-dimensional, two-dimensional and three-dimensional pore systems exist, depending on the arrangement of the channels in one, two or the three directions of the space. Examples can be found in the medium pore size zeolitic materials theta-1 (TON), ZSM-5 (MFI) and ITQ-13 (ITH) consisting of one-, two- and three-dimensional pore systems, respectively. Dimensionality of the channels is a quite important parameter for separation processes in terms of diffusion of the molecules along the porous structure. Indeed, a one-dimensional zeolite usually results a less open structure with lower micropore volume and limitations by diffusion may take place since the molecules must diffuse along one straight channel. As an example, the micropore volume of the small pore one-dimensional zeolite MCM-35 (MTF) is 0.07 cm^3 g^{-1} whereas that of the small pore three-dimensional chabasite (CHA) is 0.30 cm^3 g^{-1}.

Finally, the *channel connection* is another way of differentiating these materials and zeolites with interconnected pores, independent ones and those forming cages can be found. Examples of zeolites with interconnected pores are beta (BEA) and ZSM-5 (MFI), zeolite MCM-22 (MWW) contains independent channels and examples of zeolitic structures with cages are zeolite A (LTA) and faujasite (FAU). The presence of interconnected pores is relevant in adsorption processes since diffusion of the molecules along the channels is favoured in this type of materials. Also the presence of large cages in the structure is an important point for adsorption as these zeolites usually are characterized by very large micropore volume available for adsorption of the molecules.

The earliest reports on the use of zeolitic materials as gas adsorbents are dated on the 1950s and 1960s. Use of zeolites as adsorbents for gas chromatography able to separate gases such as hydrogen, oxygen, carbon dioxide, nitrogen, methane and carbon monoxide was reported [48]. Later, different zeolites have been described as useful adsorbents in gas mixtures containing carbon dioxide, nitrogen and methane, such as ion exchanged mordenite [49], where it is stated that the separation ability increases with increasing electrostatic field in the zeolite cavities. Also, faujasite-type zeolites (X and Y) while ion exchanged with alkali or alkaline-earth ions, NaA, CaA, silicalite, ZSM-5 and other microporous molecular sieves (*e.g.* the silicoaluminophosphate SAPO-5 and SAPO-11) have been described as adsorbents. On the other hand, natural occurring zeolites have been also used in different gas separation processes. Indeed, zeolites such as erionite, mordenite, chabasite and clinoptilolite have been widely employed as adsorbents in air separation and purification, NG purification, hydrocarbon separation, oxygen/argon and hydrogen purifications, among others [50]. A comparison on the use of three different natural zeolites (erionite, mordenite and clinoptilolite) as adsorbents for purification of methane mixed with carbon dioxide shows that clinoptilolite is the best adsorbent for purifying NG [51]. Several reports on the use of clinoptilolite exchanged with different cations as adsorbents for removing carbon dioxide and in separation of methane/nitrogen mixtures have been also reported [52] and [53]. It has been described that when using clinoptilolite for the separation of methane/nitrogen mixtures, separation efficiency can be controlled by the cationic surface population [54].

The low cost of natural zeolites is often cited as a major incentive for their use. It has been demonstrated that although the raw material cost is relatively low, processing and shipping operations could make these adsorbents even more expensive than the synthetic ones. According to that, also in the case of natural zeolites a wise evaluation of adsorbent separation performances and cost is recommended [50].

Regarding to the use of zeolites in PSA, several processes have been reported in where adsorbents such as zeolites A, X, silicalite or mordenite are employed [55], [56] and [57] and comparison with other adsorbents such as activated carbon can be also found [58]. The Petlyuk™ PSA process has been proposed for the separation of ternary gas mixtures, such as carbon dioxide/methane/nitrogen. It consists of two pairs of adsorption columns employing three kinds of adsorbents: activated carbon, zeolite 13X and carbon molecular sieves [59].

Recently, zeolite and other molecular sieve membranes have also shown potential for application in separation processes. Small-, medium-, and large-pore zeolites have been used to prepare membranes able to separate carbon dioxide from methane, as is the case of SAPO-34 [60], silicalite-1 [61] and zeolite Y [62], respectively. Because both carbon dioxide (CO_2=3.3 Å) and methane (CO_2=3.80 Å) molecules are much smaller than the pores of large pore and medium pore zeolites, separation with these membranes was mainly based on competitive adsorption. In contrast, it is reported that membranes based on small pore zeolites and molecular sieves such as zeolite DDR [63] and SAPO-34 [64], having pores similar in size to methane but larger than carbon dioxide, allow the separation of these molecules by differences in size.

A recent review on carbon dioxide separation by zeolite membranes can be found elsewhere [65]. However, it is worth to mention that the up-scaling of high quality zeolite membranes is still a major issue for commercialisation purposes. Finally, it shall be pointed out that the very recent great availability of different zeolitic materials in their pure silica composition has opened a new application field for hydrocarbon separation processes involving olefins separations. Indeed, the use of pure silica microporous materials instead of charged silicoaluminate zeolites is preferred, since the former will not suffer any pore blocking due to the oligomerization of the adsorbed olefins. In this line, non-charged zeolites with pore openings formed by 8MR, such as chabasite [66], ITQ-32 [67] and AlPO materials [68], among others, have shown excellent olefin/paraffin separation properties, suggesting separation procedures alternative to energy intensive cryogenic distillations, currently used for that purpose.

Separation of molecules by using zeolites as adsorbents can be taken place by molecular sieve effect or by selective adsorption. Regarding separation by molecular sieving and, taking into account the kinetic diameter of the molecules involved (σCO_2=3.30 Å, σN_2=3.64 Å, σCH_4=3.80 Å), zeolites with pore diameters smaller than 3.80 Å would be required. In the database of the *International Zeolite Association* (IZA) [45], 62 small pore (8MR) zeolites appear, but only four of them have pore sizes smaller than 3.80 Å (Table 7) that would be adequate for separation of carbon dioxide and nitrogen from methane.

Table 7: Small pore (8 MR) zeolites with pore size smaller than 3.8 Å

IZA Code	Name	Dimensionality	Pore Size [Å2]
ACO	ACP-1	3	2.8 × 3.5
			3.5 × 3.5
AFX	SAPO-56	3	3.4 × 3.6
	SSZ-16		
PAU	Paulingite	3	3.6 × 3.6
	ECR-18		

RHO	Rho	3	3.6 × 3.6

On the other side, the interaction of the adsorbent with the adsorbate molecules is a key factor for separation processes. Indeed, as the interaction is stronger the desorption step becomes harder and the regenerability of the adsorbent decreases. Taking into account the differences in polarity of the molecules ($CO_2 \gg N_2 \approx CH_4$), it is expected a very strong interaction of carbon dioxide with polar zeolite adsorbents making more difficult their regenerability as, indeed, has been observed in some zeolites such as CaNaA (Section 3.5). For this reason, high silica and pure silica zeolites being much less polar materials would be more desirable for this separation process provided that these adsorbents exhibit a satisfactory adsorption capacity. Among the existing small pore zeolites, those that can be synthesized in pure silica form are listed in Table 8 and could be in principle the most adequate adsorbents for this purpose.

Table 8: Small pore (8 MR) pure silica zeolites

IZA Code	Name	Dimensionality	Pore Size [Å]
CHA	Chabasite	3	3.8 × 3.8
DDR	Deca-dodecasil 3R	2	3.6 × 4.4
IHW	ITQ-32	2	3.5 × 4.3
ITE	ITQ-3	2	3.8 × 4.3
			2.7 × 5.8
ITW	ITQ-12	2	2.4 × 5.4
			3.9 × 4.2
LTA	ITQ-29	3	4.1 × 4.1
MTF	MCM-35	1	3.6 × 3.9
NSI	Nu-6(2)	2	4.5 × 2.6
			4.8 × 2.4
RTE	RUB-3	1	3.7 × 4.4
RTH	RUB-13	2	3.8 × 4.1
			2.5 × 5.6
RWR	RUB-24	1	5.0 × 2.8

Regarding to the availability for preparing and using these materials, in the case of those selected for the separation based on molecular sieve effects (Table 7), ACP-1 is a cobalt-phosphate that is unstable, paulingite is a mineral and the possible candidates would be SAPO-56, SSZ-16, ECR-18 and zeolite rho. In the case of the small pore pure silica zeolites (Table 8), most of them can be prepared and among them, it is expected that good adsorbents could be found for their use in these separation processes.

The titanosilicate ETS-4 is a small pore member of the *Engelhard TitanoSilicate* (ETS) family [69] and [70]. The ETS-4 structure consists of an interconnected octahedral–tetrahedral framework with narrow 8 MR pore openings. The ETS-4 structure involves corner sharing SiO_4 tetahedra and TiO_6 octahedra units as well as TiO_5 units. Although larger openings are present in its structure, faulting ensures that access to the crystal interior of ETS-4 occurs through the relatively narrow 8 MR, analogous to what is seen in small-pore zeolites. The size of the 8 MR pore can be tuned by exchanging cations from sodium to strontium and by dehydration using a controlled thermal treatment (between 373 K–573 K). The progressive contraction of the effective pore size of the 8 MR pore openings profoundly affects the adsorption properties. When the material is calcined at 463 K, methane is readily adsorbed while larger molecules are essentially excluded. Methane adsorption declines with further pore contraction and once the material has been dehydrated at 543 K, substantial methane exclusion occurs whereas smaller molecules (like nitrogen or carbon dioxide) are readily adsorbed. Due to these peculiar properties, an adsorbent based on ETS-4, referred as*Contracted TitanoSilicate-1* (CTS-1), has been implemented in the Molecular Gate™ technology for nitrogen rejection from NG (Section 5.1) [71] and [72].

Carbon Adsorbents

Carbon adsorbents are widely employed because of their peculiar properties, mainly due to their low polarity [73]: (i) they are able to perform separation and purification without requiring prior stringent moisture removal (in contrast to most zeolites); (ii) they adsorb more non-polar and weakly polar organic molecules than other adsorbents do; (iii) they exhibit low heat of adsorption, resulting in low energy intensive regeneration operations. Carbon adsorbents can be roughly divided into four categories:

- Activated Carbons (ACs);
- Carbon Molecular Sieves (CMS);
- Activated Carbon Fibers (ACFs);
- carbon-based nanomaterials (e.g. Single Wall Carbon Nanotubes, SWNTs).

Among them, ACs and CMS are the most employed materials in industrial gas separations.

AC source raw materials are carbonaceous matters such as wood, peat, coals, petroleum coke, bone, coconut shell, fruit nuts. AC preparative route is largely empirical although a general understanding of the related phenomena has been reported [74]. It essentially involves a low temperature (about 773 K) carbonisation followed by activation at high temperature (about 1273 K). During carbonisation, condensation of polynuclear aromatic compounds, breakage of sidechain groups together with cross-linking reactions occur. In particular, cross-linking avoids the development of graphitic structures that are virtually nonporous, thus not suitable for adsorption applications. Starting with the initial pores present in the raw material, additional pores with the desired pore size distribution can be created by activation processes. Flushing with mild oxidising gases (e.g. carbon dioxide, steam) or treatments with inorganic chemicals (e.g. potassium hydroxide, zinc chloride, phosphoric acid) are usual procedures. Both oxygenate groups and inorganic cations introduced during the activation step determine the polarity of the final product. By proper choice of precursor, carbonisation and activation steps it is possible to obtain ACs characterised by the desired pore size distribution and polarity. *Iodine number* is the most common empirical descriptor for AC adsorption capacity. It is defined as the milligrams of iodine adsorbed by one gram of carbon. Iodine number values are directly correlated with specific surface area ones.

Presently, main AC applications in the NG industry are [75]: (i) purification of recycled amines and glycols in gas sweetening and dehydration facilities; (ii) NG contaminants (e.g. sulfur, mercury) scavenging; (iii) equipment and catalyst protection. ACs impregnated with chemicals are frequently used. In this case, due to absorption phenomena, regeneration can be laborious and often promoted by high temperature heating (e.g. gas sweeping at till 723 K for alkali-impregnated ACs for sulphur scavenging from Claus reactor tail gas). Additional *ex situ* treatments can be requested.

The preparation of CMS is broadly similar but often includes additional treatments with organic species that are cracked or polymerised on the carbonised matter [76]. CMS for large scale applications are obtained from coals. A very narrow pore size distribution, in general centered on 5 Å, is obtained by accurate control of synthesis conditions. Due to precise pore size tuning, CMS are discriminated from ACs on the basis of the separation mechanism exploited. ACs separate molecules exploiting differences in their adsorption equilibrium constants. Conversely, CMS provide molecular separation on the basis of rates of adsorption. Currently, CMS are mainly employed in PSA processes for hydrogen and helium purification (which can also be done by ACs) and nitrogen production from air [77] and [78]. Carbon dioxide separation from natural gas by CMS-based PSA systems has been extensively studied [79], [80], [81], [82] and [83] and commercial units (suitable also for nitrogen rejection) are available (Sections 5.1 and 5.2).

ACFs are produced from polymeric and pitch fibers [84]. Besides their fibrous form, they have the following peculiar properties with respect to ACs: (i) narrow and uniform pore size distribution (8–10 Å) which enhance interaction with adsorbates; (ii) small and uniform fiber diameter (hence fast adsorption–desorption diffusion); (iii) electrically conductive and heat resistant allowing desorption by electrical heating; (iv) high strength and elasticity.

Despite of these favourable properties, their high cost limits the use to small units, mainly in environmental applications (e.g. air or water treatment). On the other hand, manufacture of ACF thin membranes appear promising for high purity hydrogen production from refinery gases [8] and [85].

Finally, SWNTs consist in seamless cylinders wrapped by a graphite sheet (or graphene sheet). The hexagonal honeycomb lattice of the graphene sheet can be oriented in many possible directions relative to the axis of the tube, determining the metallic or semiconducting nature of these materials. They are prepared mainly by hydrocarbon or carbon monoxide decomposition at high temperature. SWNT main characteristics are: (i) high thermal and electrical conductivities; (ii) high strengths; (iii) high stiffness. According to that, they hold potential application in high technology fields (e.g. microelectronics). As adsorbent, they have provided promising results for hydrogen storage, although careful validation of published data is recommended [86] and

[87]. Presently, neither industrial production processes nor commercial applications of these materials are reported.

Metal-Organic Frameworks

In contrast to zeolites, for which a relatively limited number of structures exists (Section 4.1), the versatility of coordination and organic chemistry allow to design an almost infinite variety of MOF structures [88], [89],[90] and [91].

MOF materials actually bridge the pore size gap between zeolite and mesostructured silica such as M41S materials, as shown in the arbitrary selection in Fig. 2. Aside from pore size and when pore topology is considered, one can find analogies between zeolites and MOFs. For instance, the pore structure can be one-dimensional with straight channels such as found for rod-like materials (MIL-53, MIL-68 and MOF-69), while IRMOF structures show three-dimensional cubic channel arrangements. In addition, as for zeolites, complex porous architecture with large cavities, reduced pore aperture and side pockets can be observed. A striking example is the HKUST-1 structure consisting of two types of "cages" and two types of "windows" separating these cages [92] and [93]. Large cages (13.2 and 11.1 Å in diameter) are interconnected by 9 Å windows of square cross-section. The large cages are also connected to tetrahedral-shaped side pockets of roughly 6 Å through triangular-shaped windows of about 4.6 Å. Typical zeolite topologies are also found in imidazolate based MOFs (also called ZIFs) such as SOD, LTA, RHO [94]. As a consequence, beta (9.6 Å) and alpha (16.4 Å) cages can be found in *sod-* and *rho-ZMOF*, respectively [95]. In terms of adsorption properties, MOF materials exhibit two major differences with respect to classical adsorbents:

- Framework flexibility upon adsorption–desorption [96] and [97]. In contrast with the permanent porosity typical of classical "rigid" adsorbents (carbons, zeolites), adsorption on MOF may evolve dynamically depending on the nature and quantity of host molecules [98], [99] and [100]. In many examples, it is observed that the adsorption process can take place stepwise at different pressures depending on the adsorbed gas leading to adsorption–desorption isotherms with hysteresis phenomena. This phenomenon, usually named *gate opening* arises mainly from the flexibility of the networks. For instance, carbon dioxide and methane adsorptions on MIL-53 are strongly affected by the presence of water which causes drastic changes in the pore shapes [101] and [102].

- Very high adsorption capacities per mass unit due to very large micro-, mesoporous volumes. This is specially the case when considering high pressure domains ($P > 25$ bar). For example, CO_2 adsorption capacities at 50 bar are twice and three time larger for MOF-177 and MIL-101 with respect to NaX [103]. These outstanding capacities can be advantageous for gas storage applications [104], [105], [106] and [88].

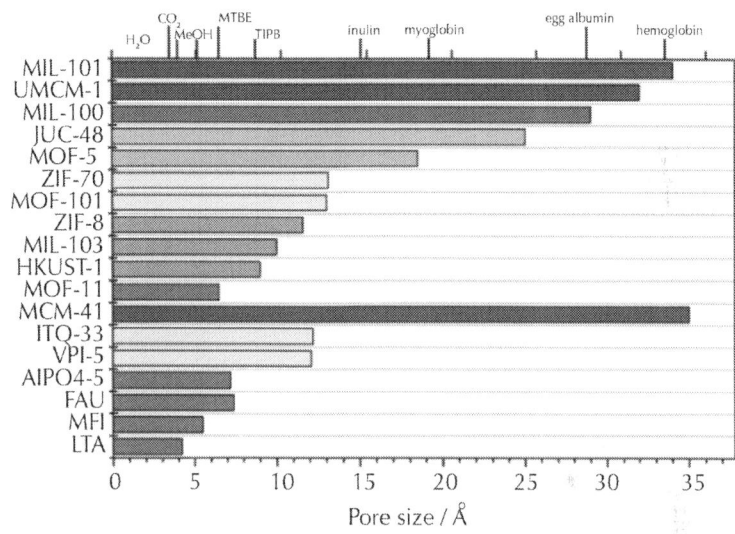

Figure 2: Cavity size of porous MOFs (Å) compared with standard aluminosilicates and aluminophosphates. Porous materials are selected arbitrarily; pore sizes are approximate due to the variety of pore shapes involved.

For comparison purposes, carbon dioxide adsorption isotherms of various carboxylate based MOFs and reference carbon materials and zeolites are plotted in Fig. 3. The stability of carboxylate based MOFs upon water adsorption is usually pointed out as a major drawback for their applications [116] and [117]. In contrast, imidazolate based MOFs (also called ZIFs) are much more chemically stable and have been recently proposed as candidates for carbon dioxide capture [94],

[118] and [119]. Similarly to high silica zeolites and carbon adsorbents, the observed linear trend of carbon dioxide isotherms on MOFs (at low pressure) is a general feature. The adsorption mainly proceeds by van der Walls and quadrupole interactions. As a result, MOF exhibits lower heats of carbon dioxide adsorption which can be taken as a decisive advantage for PSA process (Table 9). Indeed, it shall allow desorbing much more carbon dioxide at the regeneration step, thus increasing the working capacity. However, it is at the expenses of lower selectivity values which directly penalize the rate of methane recovery and which may not be sufficient to fulfil raffinate specifications (Table 2).

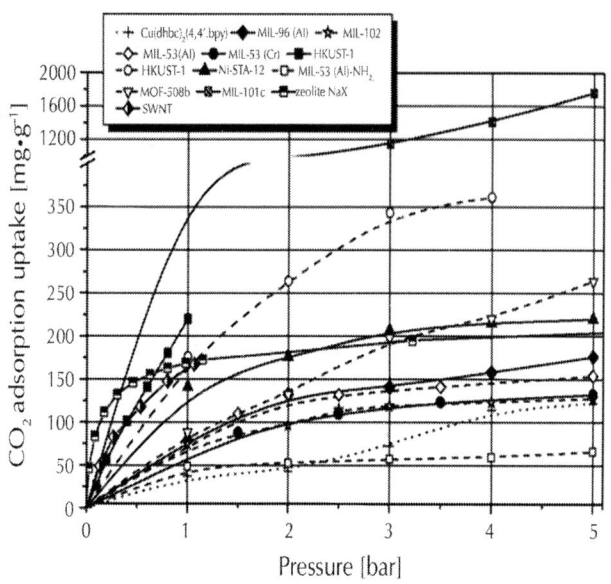

Figure 3: Carbon dioxide adsorption isotherms measured at about 300 K for various MOFs and reference materials: Cu(dhbc)$_2$(4,4'-bpy)[107], MIL-96 (Al) [108], MIL-102 [109], MIL-53 (Al) [98], MIL-53 (Cr) [101], HKUST-1 [110], HKUST-1 [111], Ni-STA-12 [99], MIL-53 (Al)-NH$_2$[112], MOF-508b [113], MIL-101c [114] and SWNTs [115].

Table 9: MOFs CO_2/CH_4 separation performaces (zeolite 13X is reported for comparison): CO_2, CH_4: Henry's selectivity, SCO_2, CH_4: selectivity estimated as the molar ratio of adsorbed carbon dioxide and methane at 5 bar, n: working capacity estimated as the difference between the adsorbed carbon dioxide at 5 bar and the adsorbed carbon dioxide at 1 bar, nCO_2: amount of adsorbed carbon dioxide at 5 bar; q_{st}: carbon dioxide isosteric heat of adsorption, N.A.: not available

Adsorbent	CO2,CH4	SCO2,CH4	n [%]	nCO2 [mg g−1]	qst [kJ mol−1]
Cu(dhbc) 2(4,4 -bpy)	N.A.	∞	64	122	N.A.
MIL-96 (Al)	N.A.	9.4	56	180	33
MIL-102	N.A.	6.5	41	128	N.A.
MIL-53 (Al)	N.A.	4.4	42	154	40
MIL-53 (Cr)	N.A.	∞	54	132	N.A.
HKUST-1	5.5	11.3	51	361	26
Ni-STA-12	N.A.	12.5	36	220	34
MOF-508b	2.9	8.2	67	264	16
MIL-101c	31	9.2	65	1760	24
NaX	93	1.8	17	208	49

Therefore new generation of MOF adsorbents shall be designed to selectively enhance carbon dioxide interactions with the framework while still limiting methane uptake. Two different strategies can be envisaged: (i) a *structural* route aiming at synthesizing MOF adsorbents with micropore size in the range of medium pore zeolites in order to benefit from stronger wall-molecules interactions. This can be achieved by networks interpenetration strategy [120]; (ii) a *functionalisation* route dealing with the enhancement of framework polarity. Simulations were performed on virtual compounds corresponding to the IRMOF topology with virtual organic linkers. It has been concluded that the use of short linkers or linkers highly functionalised with halogen moieties (e.g. bromine) would result in an increase of adsorption heats [121]. At IRCELYON, we have investigated the effect of functionalisation with basic amino groups on various carbolylate based MOFs (Fig. 4) [122] and [123]. Similarly, other authors have investigated ZIF functionalisation (ZIF-68, 78, 69, 82, 90) with polar groups such as $-NO_2$, -Cl, -CN and carboxyl[118] and [124]. Clearly, higher heats of CO_2 adsorption are found for functionalised adsorbents with respects to unpolar ZIF-8 (Table 10). However, it is worth to mention that the enhancement of the framework polarity is usually accompanied with a stronger adsorption of water as revealed by lager uptake at low pressure (Fig. 4).

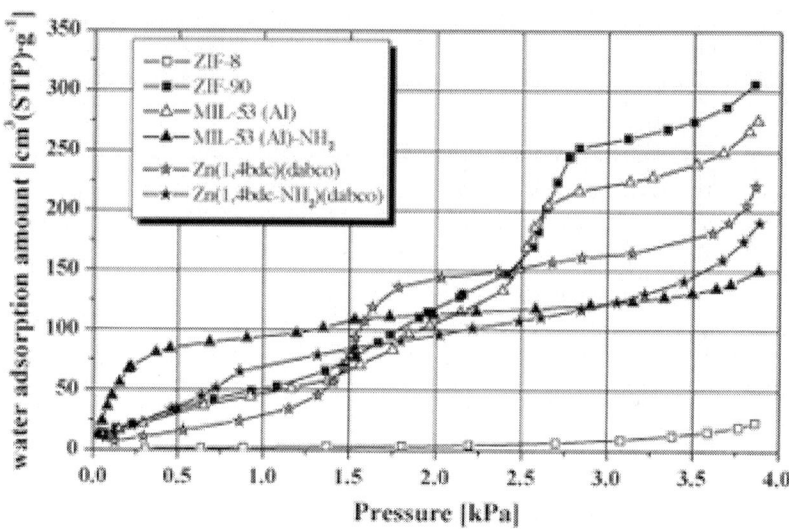

Figure 4: Effect of MOF functionalisation on water adsorption measured at 313 K.

Table 10: ZIFs CO_2/CH_4 separation performances [118] (see Table 9 for comparison)

Adsorbent	CO2,CH4	SCO2,CH4	nCO2 [mg g−1]	qst [kJ mol−1]
ZIF-68	5	3.8	72	20.8
ZIF-78	10	3.7	99	31
ZIF-82	9	4.7	101	23.9
ZIF-8	N.A.	2.8	28	N.A.

Although MOF materials are intensely investigated for capture applications, much more insights is required before envisioning any industrial application for gas separation such as the effects of mixture of various gases and vapours (including water), adsorption-desorption kinetics, shaping and poisoning issues.

PSA APPLICATIONS

PSA Applications to Nitrogen Removal from Natural Gas

The use of CMS as hydrocarbon–selective adsorbents suitable for PSA units for CBM enrichment has been suggested by Bergwerksverband (former Bergbau–Forschung) in a patent assigned in the 1980s [125]. This approach has been implemented by Nitrotec [126], [127] and [128] for nitrogen rejection. Three gas treating plants (15 MMscf d^{-1} = 4.25 × 10^5 m^3 d^{-1} each) were installed in Texas in the 1990s. Nitrotec process utilises CMS to remove hydrocarbons from the stream containing nitrogen and operates at an optimum pressure of between 2 and 4 bar. The stream flows through the CMS bed with the hydrocarbon being trapped and the nitrogen being vented. Hydrocarbons are recovered by vacuum desorption (recovery of about 95%) and thus, have to be compressed to pipeline delivery pressure.

The Nitrex™ [129] and [130] process developed by UOP by using the proprietary Polybed™ PSA platform (mainly employed for high purity hydrogen production [8]) appears to be founded on the same principles of Nitrotec one. Hydrocarbon recovery up to 95% is claimed also in this case. One nitrogen rejection unit based on this technology was commissioned in Texas in the early 1990s, with a feed flow rate of about 2.3 MMscf d^{-1} = 6.51 × 10^4 Sm^3 d^{-1}.

On the other hand, very few PSA processes for nitrogen rejection from natural gas based on nitrogen preferential adsorption have been developed. The primary issue is in finding an adsorbent that has selectivity for nitrogen over methane suitable to provide a commercial viable process. In fact, large part of known adsorbents are characterised by equilibrium adsorption selectivity that favour methane over nitrogen (Section 3.5). According to that, processes based on kinetic separation have been suggested. A TSA moving bed process for nitrogen rejection from natural gas, based on fast adsorption/desorption cycles on zeolite 4A, was patented in the 1950s [131]. The apparatus disclosed in this patent appears to be not practical and it does not provide a cost efficient separation method in view of high equipment and maintenance costs and adsorbent degradation by attrition. As a consequence, no plants based on this technology have ever been realised. On the other hand, clinoptilolite (a natural zeolite) is frequently cited in literature as rate selective adsorbent for the separation of nitrogen and methane. In particular, the use of magnesium exchanged clinoptilites has been patented by UOP [132]. Although, UOP patent indicates PSA as the preferred adsorption technology, no technical details are specified.

The Molecular Gate™ PSA process has been developed by Engelhard (now BASF Catalysts) and presently licensed by Guild Associates [18], [133], [134] and [135]. It employs a proprietary adsorbent based on the synthetic titanosilicate ETS-4 (Section 4.1). This adsorbent is referred as Contracted TitanoSilicate 1 (CTS-1). Differently to essentially all known adsorbents, CTS-1 equilibrium adsorption selectivity favours nitrogen over methane. The unique behaviour of CTS-1 adsorbent makes this process suitable for nitrogen rejection from NG by nitrogen selective adsorption. As water adsorption can hydrate and thus change the pore size of CTS-1, it is good practice to remove the water beforehand (*i.e.* by adsorption on silica gel) if present in large amount (Fig. 5).

Figure 5: Molecular Gate™ adsorption system for nitrogen and/or carbon dioxide removal (courtesy of Guild Associates, www.moleculargate.com).

Feed is introduced into Molecular Gate™ at high pressure. In most cases, optimum operating pressure is around 8 bar. Methane, ethane and about half the propane pass through the bed of adsorbent as raffinate stream, with less than 1 bar pressure drop. The system adsorbs also the residual water, all of the carbon dioxide and all of the C_{3+}. These hydrocarbons do not fit within the pore of the adsorbent. However, they are attracted to the binder used to hold the molecular sieve crystals together and are removed with the other adsorbed components into the tail gas. Adsorbent is regenerated by applying vacuum together with a minimal methane purge. The methane depleted stream (obtained as extract, at atmospheric pressure) is partially recycled to feed. Methane recovery higher than 90% has been declared.

It is interesting to point out that Molecular Gate™ technology has been proposed also for carbon dioxide rejection and successfully tested for the upgrading of biogas (containing 3000 ppmv of H_2S) to pipeline quality NG [136]. Upgrading of landfill gas contaminated by siloxanes has been also claimed [137].

Molecular Gate™ units able to treat 10 MMscf d^{-1} = 2.83 × 10^5 Sm^3 d^{-1} of NG, downsizeable till 0.5 MMscf d^{-1} = 1.41 10^4 Sm^3 d^{-1} are commercially available.

Applications to Carbon Dioxide Removal from Natural Gas

Differently to nitrogen, carbon dioxide shows great affinity for several commercial adsorbents. As a consequence, the patent literature on carbon dioxide rejection from NG by PSA is extremely rich and mainly focused on processes rather than on adsorbents.

One of the earlier patents in this field (assigned to Union Carbide, now UOP) describes a PSA system based on a zeolite able to selectively adsorb carbon dioxide from low quality NG [138]. This system exploits displacement with carbon dioxide (*rinse*) to remove adsorbed methane from zeolite bed and to purge methane from column void space. The high purity of the obtained carbon dioxide is a benefit derived by the addition of this additional step.

A similar strategy is described in a group of patents assigned to Air Products & Chemical [139],[140] and [141] in which PSA processes for carbon dioxide separation from methane (and/or hydrogen) are described. Carbon-based porous materials (ACs, CMS) are in this case the preferred adsorbents.

After desorption step, high pressure purge with carbon dioxide is followed by rinse at an intermediate pressure with an extraneous gas such as air or carbon dioxide itself. The column is then subjected to vacuum to remove the extraneous gas or any remaining carbon dioxide.

A process for methane recovery from LFG, combining a TSA and PSA units is described in a further patent assigned to Air Products & Chemicals. Specifically, the TSA unit removes water and minor impurities from the gas, which then goes to a PSA system devoted to carbon dioxide rejection [142]. The addition of a further PSA unit aimed to nitrogen rejection has been suggested to improve obtained methane quality [9].

The use of clinoptilolites for carbon dioxide rejection from poor gases is disclosed in a patent assigned to Gas Separation Technology [143]. The adsorbent has such a strong attraction to carbon dioxide that little desorption occurs even at very low pressure. According to that, regeneration by exposure to dry air is suggested.

Engelhard has successfully applied the Molecular Gate™ technology to carbon dioxide rejection from NG (Section 5.1).

Nitrotec commercialises PSA units suitable for carbon dioxide rejection from natural gas. These units operate adsorption at a pressure between 3 and 17 bar. Carbon dioxide is desorbed from the bed by vacuum [144]. In this case, the nature of the adsorbent is not specified.

BOC (now part of Linde) has presented a PSA process (based on an unspecified zeolite) for the production of fuel and carbon dioxide from LFG. According with the proposed scheme, the methane rich stream is partially used as fuel and partially converted into electricity. The carbon dioxide rich stream can be commercialised, due to its high purity. Also in this case, LFG pretreatment is suggested in order to maximise efficiency and avoid damages to the PSA unit.

The effectiveness of the BOC process is discussed as case study based on a 5 MMscf d^{-1} = 1.41 × 10^5 Sm^3 d^{-1}, feed flow [145].

Bergwerksverband (former Bergbau–Forschung) has demonstrated the effectiveness of CMS in carbon dioxide rejection from pretreated LFG. Operations have been performed during the 1980s on two pilot plants located, respectively, in Germany (850 scf d^{-1} = 2.41 · 10 Sm^3 d^{-1}, feed flow) and Holland (25 Mscf d^{-1} = 7.08 × 10^2 Sm^3 d^{-1}, feed flow). Unlike the process developed by the same company for nitrogen rejection by CMS (see Section 5.1), the inert rich stream is recovered as extract and the methane rich stream (up to 97 vol. %) as raffinate stream [146].

Osaka Gas has presented an integrated system for the exploitation of biogas: it comprises a PSA unit (for the separation of carbon dioxide from methane) and a methane adsorptive storage system. The technology (based on ACs) has been validated through a pilot plant able to treat 425 scf d^{-1} = 1.21 × 10 Sm^3 d^{-1} of biogas. Methane with the purity of 98 vol. % or higher was produced with a recovery of 90–95%. The PSA process was stable over one year [147].

CONCLUSIONS AND PERSPECTIVES

Although many adsorbents are commercially available, there are still demand for robust (high chemical stability against other contaminants, high mechanical stability against attrition), cheap (low synthesis cost since adsorbent cost represents a significant part of the investment cost) and *energy efficient* materials (e.g. with appropriate selectivity in order to limit the number of separation steps, large working capacity to reduce the cycle frequency, high regenerability to avoid use of external heat). Nevertheless, breakthrough processes can be anticipated if new materials can be design. Among others, a challenge in the materials development is the design of adsorbents which are less moisture sensitive due to the high cost of intensive drying. On the other hand, all kind of adsorbents which could process nitrogen rejection by molecular sieving shall lead to very valuable alternative processes.

Considering process engineering, *Rapid Cycle PSA* approach (RCPSA™) is a significant improvement. Xebec (former QuestAir Technologies) offers compact PSA units (0.01–9 MMscf d^{-1} = 2.83 × 10^2 to 2.54 × 10^5 Sm^3 d^{-1} of purified NG), based on proprietary rotary valves [148] and [149] and state of the art adsorbents (mainly for carbon dioxide and water removal). Xebec plans to build larger capacity PSA systems for NG processing based on the rapid cycle RCPSA™ system jointly developed with ExxonMobil Research and Engineering [150] and [151]. RCPSA exploits both proprietary valves and structured adsorbents. Structured adsorbents overcome fluidisation limitations of beaded adsorbents, allowing higher cycle speeds (up to 50 cycles per minute) than conventional systems. The result is a significant size reduction of separation units (till 1/20 of state of the art PSA plants of the same productivity). Furthermore, the rotary valve technology replaces the bulky network of piping and valves used in conventional PSA systems with two compact, integrated valves. Structured adsorbent and rotary valve are packaged into modules as depicted in Fig. 6. Modular designs are being developed with total capacities up to 80 MMScf d^{-1} = 2.27×10^6 Sm^3 d^{-1} [152]. A prototype has been operated at an ExxonMobil refinery for the production of high purity hydrogen (Fig. 7).

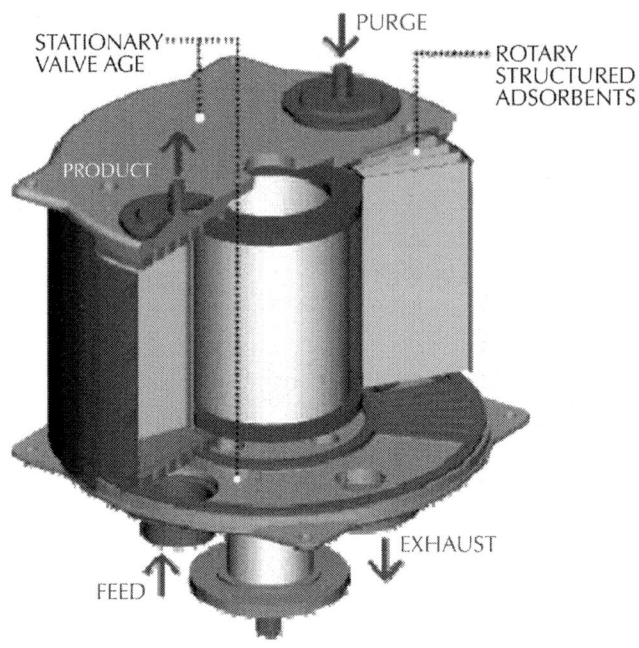

Figure 6: RCPSA™ module scheme (courtesy of Xebec, www.xebecinc.com).

Figure 7: Refinery application of RCPSA™ technology (courtesy of Xebec, www.xebecinc.com).

ACKNOWLEDGEMENTS

This work was supported by TopCombi (NMP2-CT2005-515792) funded by the European Union and by MECAFI (ANR-07PCO2-003-05), ACACIA (ANR-08PCO2-001), funded by the *French Research National Agency* (ANR). The authors thank the colleagues Enrico F. Gambarotta, Angelo Carugati, Francesco Frigerio (all of Eni), Michael J. Mitariten (Guild Associates), Daryl Musselman (Xebec), Fernando Rey (Instituto de Tecnología Química, Valencia) for the effective cooperation.

REFERENCES

1. G.E. Keller II, Industrial gas separation, in: T.E. Whyte Jr., C.M. Yon, E.H. Wagener (Eds.), ACS Symposia Series, vol. 223, ACS, Washington, USA, 1983.

2. International Adsorption Society, 2008, http://ias.vub.ac.be.

3. D.M. Ruthven, S. Farouw, K.S. Knaebel, Pressure Swing Adsorption, WileyVCH, New York, USA, 1994.

4. D.M. Ruthven, Principles of Adsorption and Adsorption Processes, Wiley, New York, USA, 1984.

5. J.L. Humphrey, G.E. Keller, Separation Process Technology, McGraw-Hill, New York, USA, 1997.

6. R.T. Yang, Adsorbents: Fundamentals and Applications, Wiley, New York, USA, 2003.

7. S. Sircar, Pressure swing adsorption, Ind. Eng. Chem. Res. 41 (2002) 1389–1392.

8. S. Sircar, T.C. Golden, Purification of hydrogen by pressure swing adsorption, Sep. Sci. Technol. 35 (2000) 667–687.

9. Knaebel, D. Ruthven, J.L. Humphrey, R. Carr, in: P.P. Radecki, J.C. Crittenden, D.R. Shonnard, J.L. Bulloch (Eds.), Emerging Separation and Separative Reaction Technologies for Process Waste Reduction: Adsorption and Membrane Systems, AIChE Center for Waste Reduction Technologies, New York, USA, 1999.

10. Eni sustainability report, 2006, www.eni.it.

11. Eni world oil & gas review, 2007, www.eni.it.

12. A.J. Kidnay, W.R. Parrish, Fundamentals of Natural Gas Processing, Taylor & Francis, Boca Raton, USA, 2006.

13. Gas Processors Suppliers Association Engineering Data Book, 12th ed., GPSA, Tulsa, USA, 2004.

14. R.M. Flores, Coalbed methane: from hazard to resource, Int. J. Coal Geol. 35 (1998) 3–26.

15. G. Nagl, Recover high-Btu gas from new sources, Hydrocarbon Process. 86 (2007) 45–48.

16. A. Rojey, C. Jaffret, S. Cornot-Gandolphe, B. Durand, S. Jullian, M. Valais, Natural Gas: Production Processing Transport, Editions Technip, Paris, France, 1997.

17. A. Kohl, R. Nielsen, Gas Purification, 5th ed., Gulf Publishing, Houston, USA, 1997.

18. U. Daimiger, W. Lind, Adsorption processes for natural gas treatment: a technology update, Engelhard Brochure (2004).

19. M.A. Huffmaster, Gas dehydration fundamentals: introduction, in: Proceedings of the 54th Laurance Reid Gas Conditioning Conference, Norman, USA, 2004.

20. K. Pale, K. Lokhandwala, Advances in membrane materials provide new solutions in the gas business, in: Proceedings of the 54th Laurance Reid Gas Conditioning Conference, Norman, USA, 2004.

21. J. Johnson, Chem. Eng. News 85 (2007) 10–110.

22. C.W. Skarstrom, Method and apparatus for fractionating gaseous mixtures by adsorption, US Patent 2,944, 627 (1960) assigned to Esso Research & Engineering.

23. P. Guerin de Motgareuil, D. Domine, Procédé de séparation d'un mélange gazeux binarie par adsorption, French Patent 1,223,261 (1960) assigned to L'Air Liquide.

24. S.J. Gregg, K.S.W. Sing, Adsorption, Surface Area and Porosity, Academic Press, London, Great Britain, 1982.

25. F. Roquerol, J. Roquerol, K.S.W. Sing, Adsorption by Powders and Porous Solids: Principles, Methodology and Applications, Academic Press, London, Great Britain, 1999.

26. D.W. Breck, Zeolite Molecular Sieves, John Willey & Sons, New York, USA, 1974.

27. NIST Computational chemistry comparison and benchmark database, NIST standard reference database n. 101 rel. 14, September 2006, in: R.D. Johnson III (Ed.), http://srdata.nist.gov/cccbdb.

28. R.H. Perry, D.W. Green, Perry's Chemical Engineers' Handbook, McGraw-Hill, New York, USA, 1999.

29. G.H. Kühl, in: J. Weitkamp, L. Puppe (Eds.), Catalysis and Zeolites: Fundamentals and Applications, Springer, New York, USA, 1999.

30. H.P. Boehm, Surface oxides on carbon and their analysis: a critical assessment, Carbon 40 (2002) 145–149.

31. K. Kaneko, Micropore filling mechanism in inorganic sorbents, in: A. Dabrowski, V.A. Tertykh (Eds.), Studies in Surface and Catalysis, Elsevier, Amsterdam, 1996, pp. 573–598.

32. M.M. Dubinin, The potential theory of adsorption of gases and vapors for adsorbents with energetically nonuniform surfaces, Chem. Rev. 60 (1960) 235–241.

33. S. Brunauer, P.H. Emmett, E. Teller, Adsorption of gases in multimolecular layers, J. Am. Chem. Soc. 60 (1938) 309–319.

34. A. Macario, A. Katovic, G. Giordano, F. Lucolano, D. Caputo, Synthesis of mesoporous materials for carbon dioxide sequestration, Microporous Mesoporous Mater. 81 (2005) 139–147.

35.   D.P. Valezuela, A.L. Myers, Adsorption Equilibrium Data Handbook, Prentice Hall, Englewood Cliffs, USA, 1989.

36.   S. Sircar, Separation of methane and carbon dioxide gas mixtures by pressure swing adsorption, in: Proceedings of the 78th AIChE Annual Meeting, New York, USA, 1987 (paper 100c).

37.   S. Himeno, T. Komatsu, S. Fujita, High-pressure adsorption equilibria of methane and carbon dioxide on several activated carbons, J. Chem. Eng. Data 50 (2005) 369–376.

38.   J.A. Dunne, R. Mariwala, M. Rao, S. Sircar, R.J. Gorte, A.L. Myers, Calorimetric heats of adsorption and adsorption isotherms. 1. O2, N2, Ar, CO2, CH4, C2H6, and SF6 on Silicalite, Langmuir 12 (1996) 5888–5895.

39.   J.A. Dunne,M. Rao, S. Sircar, R.J. Gorte, A.L.Myers, Calorimetric heats of adsorption and adsorption isotherms. 2. O2, N2, Ar, CO2, CH4, C2H6, and SF6 on NaX, H-ZSM-5, and Na-ZSM-5 zeolites, Langmuir 12 (1996) 5896–5904.

40.   S.U. Rege, R.T. Yang, M.A. Buzanowski, Sorbents for air prepurification in air separation, Chem. Eng. Sci. 55 (2000) 4827–4838.

41.   A. Jayaraman, A.J. Hernandez-Maldonado, R.T. Yang, D. Chinn, C.L. Munson, D.H. Mohr, Clinoptilolites for nitrogen/methane separation, Chem. Eng. Sci. 59 (2004) 2407–2417.

42.   M.W. Ackley, Multilayer adsorbent for PSA gas separation, US 6,152,991 (2000) assigned to Praxair.

43.   S.U. Rege, R.T. Yang, A simple parameter for selecting an adsorbent for gas separation by pressure swing adsorption, Sep. Sci. Technol. 36 (2001) 3355–3365.

44.   J. Kärger, D.M. Ruthven, Diffusion in Zeolites, Wiley, New York, USA, 1992.

45.   International Zeolite Association database of zeolite structures, http://www.iza-structure.org/databases.

46.   A. Corma, From microporous to mesoporous molecular sieve materials and their use in catalysis, Chem. Rev. 97 (1997) 2373–2419.

47.   R.M. Barrer, Zeolites and Clay Materials as Sorbents and Molecular Sieves,\ Academic Press, London, Great Britain, 1978.

48. J. Janak, M. Kejci, H.E. Dubsky, Properties of Ca zeolite as an adsorbent for gas chromatography, Ann. N. Y. Acad. Sci. 72 (1959) 731–738.

49. E.F. Vansant, R. Voets, Adsorption of binary gas-mixtures in ion-exchanged forms of mordenite, J. Chem. Soc. Faraday Trans. I 77 (1981) 1371–1380.

50. M.W. Ackley, S.U. Rege, H. Saxena, Application of natural zeolites in the purifi- cation and separation of gases, Microporous Mesoporous Mater. 61 (2003) 25–42.

51. R. Hernandez-Huesca, L. Diaz, G. Aguilar-Armenta, Adsorption equilibria and kinetics of CO2, CH4 and N2 in natural zeolites, Sep. Purif. Technol. 15 (1999) 163–173.

52. C.C. Chao, H. Rastelli, Process for purification of hydrocarbons using metal exchanged clinoptilolite to remove carbon dioxide, US Patent 4,935,580 (1990) assigned to UOP.

53. M.W. Ackley, R.T. Yang, Adsorption characteristics of high-exchange clinoptilolites, Ind. Eng. Chem. Res. 30 (1991) 2523–2530.

54. T.C. Frankiewicz, R.G. Donnelly, in: T.E.White (Ed.), Industrial Gas Separations, vol. 11, ACS, Washington, USA, 1983.

55. S. Sircar, Regeneration of adsorbents, US Patent 4,784,672 (1988) assigned to Air Products & Chemicals.

56. E. Richter, Industrial processes for gas separation by pressure swing adsorption, Erdol und Kohle. Erdgas. Petrochemie 40 (1987) 432–438.

57. H. Rastelli, C.C. Chao, D.R. Garg, Selective adsorption of CO2 on zeolite, US Patent 4,775,396, (1988) assigned to Union Carbide.

58. J. Hart, W. Thomas, Gas separation by pulsed pressure swing adsorption, Gas Sep. Purif. 5 (1991) 125–133.

59. F. Dong, H.M. Lou, A. Kodama, M. Goto, T. Hirose, The Petlyuk PSA process for the separation of ternary gas mixtures: exemplification by separating a mixture of CO2-CH4-N2, Sep. Purif. Technol. 16 (1999) 159–166.

60. J.C. Poshusta, V.A. Tuan, E.A. Pape, R.D. Noble, J.L. Falconer, Separation of light gas mixtures using SAPO-34 membranes, AIChE J. 46 (2000) 779–789.

61. L.J.P. van den Broeke, F. Kapteijn, J.A. Moulijn, Transport and separation properties of a silicalite-1 membrane - II. Variable separation factor, Chem. Eng. Sci. 54 (1999) 259–269.

62. K. Kusakabe, T. Kuroda, A. Murata, S. Morooka, Formation of a Y-type zeolite membrane on a porous alpha-alumina tube for gas separation, Ind. Eng. Chem. Res. 36 (1997) 649–655.

63. T. Tomita, K. Nakayama, H. Sakai, Gas separation characteristics of DDR type zeolite membrane, Microporous Mesoporous Mater. 68 (2004) 71–75.

64. S.G. Li, J.G. Martinek, J.L. Falconer, R.D. Noble, T.Q. Gardner, High-pressure CO2/CH4 separation using SAPO-34 membranes, Ind. Eng. Chem. Res. 44 (2005) 3220–3228.

65. M. Pera-Titus, S. Miachon, J. Sublet, S. Aguado, D. Farrusseng, Ceramic membranes for CO2 capture: present and prospects, submitted for publication.

66. D. Olson, Light hydrocarbon separation using eight-member ring zeolites, US Patent 6,488,741 (2002) assigned to The Trustess of the University of Pennsylvania.

67. M. Palomino, A. Cantin, A. Corma, S. Leiva, F. Rey, S. Valencia, Pure silica ITQ- 32 zeolite allows separation of linear olefins from paraffins, Chem. Commun. (2007) 1233–1235.

68. S. Reyes, V. Krishnan, G. DeMartin, J. Sinflet, K. Strohmaier J. Santiesteban, Separation of propylene from hydrocarbon mixtures, World Patent 03/080548 (2003) assigned to ExxonMobil Research and Engineering.

69. S.M. Kuznicki, V.A. Bell, S. Nair, H.W. Hillhouse, R.M. Jacubinas, C.M. Braunbarth, B.H. Toby, M. Tsapatsis, A titanosilicate molecular sieve with adjustable pores for size-selective adsorption of molecules, Nature 412 (2001) 720– 724.

70. R.P. Marathe, K. Mantri, M.P. Srinivasan, S. Farooq, Effect of ion exchange and dehydration temperature on the adsorption and diffusion of gases in ETS-4, Ind. Eng. Chem. Res. 43 (2004) 5281– 5290.

71. S.M. Kuznicki, V.A. Bell, I. Petrovic, P. Blosser, Separation of nitrogen from mixtures thereof with methane utilising barium exchanged ETS-4, US Patent 5,989,316 (2000) assigned to Engelhard.

72. S.M. Kuznicki, V.A. Bell, I. Petrovic, B.T. Desai, Small pored crystalline titanium molecular sieve zeolites and their use in gas separation processes, US Patent 6,068,682 (2000) assigned to Engelhard.

73. H. Juntgen, New applications for carbonaceous adsorbents, Carbon 15 (1977) 273–283.

74. T.J. Barton, L.M. Bull, W.G. Klemperer, D.A. Loy, B. McEnaney, M. Misono, P.A. Monson, G. Pez, G.W. Scherer, J.C. Vartuli, O.M. Yaghi, Tailored porous materials, Chem. Mater. 11 (1999) 2633–2656.

75. M.J. Bourke, A.F. Mazzoni, The roles of activated carbons in gas conditioning, in: Proceedings of the 39th Laurance Reid Gas Conditioning Conference, Norman, USA, 1989.

76. H.C. Foley, Carbogenic molecular-sieves—Synthesis, properties and applications, Microporous Mater. 4 (1995) 407–433.

77. S. Sircar, T.C. Golden, M.B. Rao, Activated carbon for gas separation and storage, Carbon 34 (1996) 1–12.

78. H. Juntgen, K. Knoblauch, K. Harder, Carbon molecular-sieves— Production from coal and application in gas separation, Fuel 60 (1981) 817–822.

79. A. Jayaraman, A.S. Chiao, J. Padin, R.T. Yang, C.L. Munson, Kinetic separation of methane/carbon dioxide by molecular sieve carbons, Sep. Sci. Technol. 37 (2002) 2505–2528.

80. A. Kapoor, R.T. Yang, Kinetic separation of methane carbon dioxide mixture by adsorption on molecular-sieve carbon, Chem. Eng. Sci. 44 (1989) 1723– 1733.

81. A. Kapoor, R.T. Yang, C. Wong, Surface diffusion, Catal. Rev. Sci. Eng. 31 (1989) 129–214.

82. V.G. Gomes, M.M. Hassan, Coalseam methane recovery by vacuum swing adsorption, Sep. Purif. Technol. 24 (2001) 189–196.

83. S. Cavenati, C.A. Grande, A.E. Rodrigues, Upgrade of methane from landfill gas by pressure swing adsorption, Energy Fuels 19 (2005) 2545–2555.

84. M. Suzuki, Activated carbon fiber: fundamentals and applications, Carbon 32 (1994) 577–586.

85. S. Sircar, W.E. Waldron, M. Anand, M.D. Rao, Hydrogen recovery by pressure swing adsorption integrated with adsorbent membranes, US Patent 5,753,010 (1998) assigned to Air Products and Chemicals.

86. H.G. Schimmel, G.J. Kearley, M.G. Nijkamp, C.T. Visserl, K.P. de Jong, F.M. Mulder, Hydrogen adsorption in carbon nanostructures: comparison of nanotubes, fibers, and coals, Chem. Eur. J. 9 (2003) 4764–4770.

87. C. Zandonella, Is it all just a pipe dream? Nature 410 (2001) 734–735.

88. G. Ferey, Hybrid porous solids: past, present, future, Chem. Soc. Rev. 37 (2008) 191–214.

89. C. Mellot-Draznieks, G. Ferey, Assembling molecular species into 3D frameworks: computational design and structure solution oh hybrid materials, Prog. Solid State Chem. 33 (2005) 187–197.

90. S. Natarajan, S. Mandal, Open-framework structures of transition-metal compounds, Angew. Chem. Int. Ed. 47 (2008) 4798–4828.

91. D.J. Tranchemontagne, J.L. Mendoza-Corteʾis, M. O'Keeffe, O.M. Yaghi, Secondary building units, nets and bonding in the chemistry of metal–organic frameworksw, Chem. Soc. Rev. 38 (2009) 1257–1283.

92. C. Chmelik, J. Kaerger, M. Wiebcke, J. Caro, J.M. van Baten, R. Krishna, Adsorption and diffusionof alkanes in CuBTC crystals investigated using infra-red microscopy and molecular simulations, Microporous Mesoporous Mater. 117 (2008) 22–32.

93. J.L.C. Rowsell, O.M. Yaghi, Effects of functionalization, catenation, and variation of the metal oxide and organic linking units on the low-pressure hydrogen adsorption properties of metal-organic frameworks, J. Am. Chem. Soc. 128 (2006) 1304–1315.

94. R. Banerjee, A. Phan, B. Wang, C. Knobler, H. Furukawa, M. O'Keeffe, O.M. Yaghi, High-throughput synthesis of zeolitic imidazolate frameworks and application to CO2 capture, Science 319 (2008) 939–943.

95. Y. Liu, V.C. Kravtsov, R. Larsen, M. Eddaoudi, Molecular building blocks approach to the assembly of zeolite-like metal-organic frameworks (ZMOFs) with extra-large cavities, Chem. Commun. 14 (2006) 1488–1490.

96. A.J. Fletcher, K.M. Thomas, M.J. Rosseinsky, Flexibility in metal-organic framework materials: impact on sorption properties, J. Solid State Chem. 178 (2005) 2491–2510.

97. K. Uemura, S. Kitagawa, M. Kondo, K. Fukui, R. Kitaura, H.C. Chang, T. Mizutani, Novel flexible frameworks of porous cobalt(III) coordination polymers that show selective guest adsorption based on the switching of hydrogen-bond pairs of amide groups, Chem. Eur. J. 8 (2002) 3586–3600.

98. S. Bourrelly, P.L. Llewellyn, C. Serre, F. Millange, T. Loiseau, G. Ferey, Different adsorption behaviors of methane and carbon dioxide in the isotypic nanoporous metal terephthalates MIL-53 and MIL-47, J. Am. Chem. Soc. 127 (2005) 13519–13521.

99. S.R. Miller, G.M. Pearce, P.A. Wright, F. Bonino, S. Chavan, S. Bordiga, I. Margiolaki, N. Guillou, G. Ferey, S. Bourrelly, P.L. Llewellyn, Structural transformations and adsorption of fuel-related gases of a structurally responsive nickel phosphonate metal-organic framework, Ni-STA-12, J. Am. Chem. Soc. 130 (2008) 15967–15981.

100. S. Kitagawa, R. Kitaura, S. Noro, Functional porous coordination polymers, Angew. Chem. Int. Ed. 43 (2004) 2334–2375.

101. P.L. Llewellyn, S. Bourrelly, C. Serre, Y. Filinchuk, G. Ferey, How hydration drastically improves adsorption selectivity for CO2 over CH4 in the flexible chromium terephthalate MIL-53, Angew. Chem. Int. Ed. 45 (2006) 7751–7754.

102. T. Loiseau, C. Serre, C. Huguenard, G. Fink, F. Taulelle, M. Henry, T. Bataille, G. Ferey, A rationale for the large breathing of the porous aluminum terephtalate (MIL-53) upon hydratation, Chem. Eur. J. 10 (2004) 1373–1382.

103. A. Millward, O.M. Yaghi, Metal-organic frameworks with expeptionally high capacity for storage of carbon dioxide at room temperature, J. Am. Chem. Soc. 127 (2005) 17998–17999.

104. R. Morris, P. Wheatley, Gas storage in nanoporous materials, Angew. Chem. Int. Ed. 47 (2008) 4966–4981.

105. U. Mueller, M. Schubert, F. Teich, H. Puetter, K. Schierle-Arndt, J. Pastre, Metalorganic frameworks—prospective industrial applications, J. Mater. Chem. 16 (2006) 626–636.

106. J.R. Li, R.J. Kuppler, H.C. Zhou, Selective gas adsorption and separation in metal-organic frameworks, Chem. Soc. Rev. 38 (2009) 1477–1504.

107. R. Kitaura, K. Seki, G. Akiyama, S. Kitagawa, Porous coordination-polymer crystals with gated channels specific for supercritical gases, Angew. Chem. Int. Ed. 42 (2003) 428–431.

108. T. Loiseau, L. Lecroq, C. Volkringer, J. Marrot, G. Ferey, M. Haouas, F. Taulelle, S. Bourrelly, P.L. Llewellyn, M. Latroche, MIL-96, a Porous aluminum trimesate 3D structure constructed from a hexagonal network of 18-membered rings and m3-oxo-centered trinuclear units, J. Am. Chem. Soc. 128 (2006) 10223–10230.

109. S. Surble, F. Millange, C. Serre, T. Duren, M. Latroche, S. Bourrelly, P.L. Llewellyn, G. Ferey, Synthesis of MIL-102, a chromium carboxylate metalorganic framework, with gas sorption analysis, J. Am. Chem. Soc. 128 (2006) 14889–14896.

110. Q.M. Wang, D. Shen, M. Bulow, M.L.L. Lau, S. Deng, F.R. Fitch, N.O. Lemcoff, J. Semanscin, Metallo-organic molecular sieve for gas separation and purification, Microporous Mesoporous Mater. 55 (2002) 217–230.

111. S. Cavenati, C.A. Grande, A.E. Rodrigues, C. Kiener, U. Mu$\hat{i}$ller, Metal organic framework adsorbent for biogas upgrading, Ind. Eng. Chem. Res. 47 (2008) 6333–6335.

112. S. Couck, J.F.M. Denayer, G.V. Baron, J. Gascon, F. Kapteijn, An aminefunctionalized MIL-53 metal-organic framework with large separation power for CO_2 and CH_4, J. Am. Chem. Soc. 131 (2009) 6326–6327.

113. L. Bastin, P.S. Barcia, E.J. Hurtado, J.A.C. Silva, A.E. Rodrigues, B. Chen, A microporous metal-organic framework for separation of CO_2/N_2 and CO_2/CH_4 by fixed-bed adsorption, J. Phys. Chem. C 112 (2008) 1575–1581.

114. P.L. Llewellyn, S. Bourrelly, C. Serre, A. Vimont, M. Daturi, L. Hamon, G. De Weireld, J.-S. Chang, D.-Y. Hong, Y. Kyu Hwang, S. Hwa Jhung, G. Ferey, High uptakes of CO_2 and CH_4 in mesoporous Metal-organic frameworks MIL-100 and MIL-101, Langmuir 24 (2008) 7245–7250.

115. M. Cinke, J. Li, C.W. Bauschlicher, A. Ricca, M. Meyyappan, CO2 adsorption in single-walled carbon nanotubes, Chem. Phys. Lett. 376 (2003) 761–766.

116. S. Kaye, A. Dailly, O.M. Yaghi, J.R. Long, Impact of preparation and handling on the hydrogen storage properties of ZnO(1,4-benzenedicarboxylate) (MOF-5), J. Am. Chem. Soc. 26 (2007) 14176–14177.

117. P. Küsgens, M. Rose, I. Senkovska, H. Fröde, A. Henschel, S. Siegle, S. Kaskel, Characterization of metal-organic frameworks by water adsorption, Microporous Mesoporous Mater. 120 (2009) 325–330.

118. R. Banerjee, H. Furukawa, D. Britt, C. Knobler, M. OKeeffe, O.M. Yaghi, Control of pore size and functionality in isoreticular zeolitic imidazolate frameworks and their carbon dioxide selective papture properties, J. Am. Chem. Soc. 131 (2009) 3875–3877.

119. H. Hayashi, A.P. Cote, H. Furukawa, M. O'Keeffe, O.M. Yaghi, Zeolite: A imidazolate frameworks, Nat. Mater. 6 (2007) 501–506.

120. M. Eddaoudi, J. Kim, N. Rosi, D. Vodak, J. Wachter, M. O'Keefe, O.M. Yaghi, Systematic design of pore size and functionality in isoreticular MOFs and their application in methane storage, Science 295 (2002) 469–472.

121. T. Duren, F. Millange, G. Ferey, K.S. Walton, R.Q. Snurr, Calculating geometric surface aeras as a characterization tool for Metal-Organic Frameworks, J. Phys. Chem. C 111 (2007) 15350–15356.

122. D. Farrusseng, C. Mirodatos, Rational design and synthesis of Metal Open Frameworks for adsorption and catalysis, in: U.S. Ozkan (Ed.), Handbook of Heterogeneous Catalyst Design, Wiley-VCH, Weiheim, 2009, pp. 161–190.

123. D. Farrusseng, C. Daniel, C. Gaudilleìre, U. Ravon, Y. Schuurman, C. Mirodatos, D. Dubbeldam, H. Frost, R.Q. Snurr, Heats of adsorption for seven gases in three metal organic frameworks: systematic comparison of experiment and simulation, Langmuir 25 (2009) 7383–7388.

124. W. Morris, C.J. Doonan, H. Furukawa, R. Banerjee, O.M. Yaghi, Crystals as molecules: postsynthesis covalent functionalization of

zeolitic imidazolate frameworks, J. Am. Chem. Soc. 130 (2008) 12626–12627.

125. E. Richter, W. Körnbächer, K. Knoblauch, K. Giessler, Method of producing a methane-rich gas mixture from mine gas, US Patent 4,521,221 (1985) assigned to Bergwerksverband.

126. H.E. Reinhold III, J.S. D'amico, K.S. Knaebel, Natural gas enrichment process, US Patent 5,536,300 (1996) assigned to Nitrotec.

127. Gas Reseach Institute Topical Report: Technologies for nitrogen removal, GRI- 99/0080, GRI, Chicago, USA, May 1999.

128. PSA nitrogen rejection technology, 2005, www.nitrotec.com.

129. Nitrogen rejection: NitrexTM, Hydrocarbon Processing, April Issue, 1998, p. 116.

130. R.J. Buras, M.J. Mitariten, Nitrogen rejection with pressure swing adsorption: principles, design and remote control using an expert system, in: Proceed- ings of the 44th Laurance Reid Gas Conditioning Conference, Norman, USA, 1994.

131. H.W. Habgood, Removal of nitrogen from natural gas, US Patent 2,843,219 (1958) assigned to Canadian Patents and Development.

132. C.C. Chao, Selective adsorption on magnesium-containing clinoptilolite, US Patent 4,964,889 (1990) assigned to UOP.

133. K.F. Butwell,W.B. Dolan S.M. Kuznicki, Selective removal of nitrogen from natural gas by pressure swing adsorption, US Patent 6,197,092 (2001) assigned to Engelhard.

134. Nitrogen Rejection and Carbon Dioxide Removal Made Easy, Guild Associates brochure, 2008. www.moleculargate.com.

135. Carbon dioxide removal: Molecular GateTM, Hydrocarbon Processing Gas Processes 2006, Gulf Publishing, Houston, USA, 2006.

136. Developments in gas separation, Guild Associates Brochure, 2007, www.moleculargate.com.

137. M.J. Mitariten, Landfill gas upgrading process, US Patent Application 2007/0068386 (2007) submitted by Engelhard.

138. J.J. Collins, Bulk separation of carbon dioxide from natural gas, US Patent 3,751,878 (1973) assigned to Union Carbide.

139. S. Sircar, J.W. Zondlo, Hydrogen purification by selective adsorption US Patent 4,077,779 (1978) assigned to Air Products and Chemicals.

140. S.P. DiMartino, Vacuum swing adsorption process with vacuum aided internal rinse, US Patent 4,857,083 (1989) assigned to Air Products and Chemicals.

141. R. Kumar, Adsorptive process for producing two gas streams from a gas mixture, US Patent 4,915,711 (1990) assigned to Air Products and Chemicals.

142. R. Kumar, Adsorptive process for producing two gas streams from a gas mixture, US Patent 5,026,406 (1990) assigned to Air Products and Chemicals.

143. M.W. Seery, Bulk separation of carbon dioxide from methane using natural clinoptilolite US Patent 5,938,819 (1999) assigned to Gas Separation Technology.

144. Carbon dioxide rejection technology, 2005, www.nitrotec.com.

145. V.A. Malik, S.L. Lerner, D.L. MacLean, Electricity, CH_4 and Liquid CO_2 production from landfill gas, Gas Sep. Purif. 1 (1987) 77–83.

146. E. Pilarczyk, K.D. Henning, K. Knoblauch, Natural gas from landfill gases, Resourc. Conserv. 14 (1987) 283–294.

147. K. Seki, K. Mochidzuki, A. Sakoda, Adsorptive separation and storage of methane from biogas, in: Proceedings of the 9th International Conference on Fundamental of Adsorption, Giardini Naxos, Italy, 2007.

148. B.G. Keefer, D.G. Doman, Flow regulated pressure swing adsorption system, US Patent 6,063,161 (2000) assigned to SoFinoy Societte Financiere d'Innovation.

149. QuestAir M3100/M3200TM methane purification PSA systems, QuestAir Technologies Brochure, 2002.

150. S. Alizadeh-Khiavi, J.A. Sawada, A.C. Gibbs, J. Alvaji, Rapid cycle syngas pressure swing adsorption system, US Patent Application 2007/0125228 (2007) submitted by QuestAir Technologies.

151. Pressure swing adsorption: rapid cycle, Hydrocarbon Processing Refining Processes 2006, Gulf Publishing, Houston, USA, 2006.

152. A. Sapre, J. Poturovic, A. Wanni, T. Melli, ExxonMobil advanced technologies: refiners solution to present and future industry challenges, in: Proceedings of the 8th International Downstream Technology Conference and Exhibition, London, Great Britain, 2007.

Citations

CHAPTER 1

A. A. Rabah and S. A. Mohamed, "Prediction of Molar Volumes of the Sudanese Reservoir Fluids," Journal of Thermodynamics, vol. 2010, Article ID 142475, 9 pages, 2010. doi:10.1155/2010/142475.

CHAPTER 2

Karaei, M. , Ahmadi, A. , Fallah, H. , Kashkooli, S. and Bahmanbeglo, J. (2015) Field Scale Simulation Study of Miscible Water Alternating CO2 Injection Process in Fractured Reservoirs. Geomaterials, 5, 25-33. doi:10.4236/gm.2015.51003.

CHAPTER 3

Rustem Zaydullin, Denis V. Voskov, Scott C. James, Heath Henley, and Angelo Lucia, Fully Compositional and Thermal Reservoir Simulation, doi:10.1016/j.compchemeng.2013.12.008.

CHAPTER 4

A. Asgari, M. Dianatirad, M. Ranjbaran A.R. Sadeghi and M.R. Rahimpour, Methanol treatment in gas condensate reservoirs: A modeling and experimental study, doi:10.1016/j.cherd.2013.08.015.

CHAPTER 5

Mahendra P. Verma, GeoSys.Chem: Estimate of Reservoir Fluid Characteristics as First Step in Geochemical Modeling of Geothermal Systems, doi:10.1016/j.cageo.2012.06.001.

CHAPTER 6

Zhenzi Jing, Kimio Watanabe, onathan Willis-Richards, and Toshiyuki Hashida, A 3-D Water/Rock Chemical Interaction Model For Prediction Of HDR/HWR Geothermal Reservoir Performance, doi:10.1016/S0375-6505(00)00059-6.

CHAPTER 7

A. Stanton and A.A. Javadi, An automated approach for an optimised least cost solution of reinforced concrete reservoirs using site parameters, doi:10.1016/j.engstruct.2013.12.020.

CHAPTER 8

Marco Tagliabue, David Farrusseng, Susana Valencia, Sonia Aguado, Ugo Ravon, Caterina Rizzo, Avelino Corma, and Claude Mirodatos, Natural Gas Treating By Selective Adsorption: Material Science and Chemical Engineering Interplay, doi:10.1016/j.cej.2009.09.010

Index

A

Acid-neutralizing capacity (ANC) 119
Adachi-Lu-Sugie (ALS) 1, 2
Automatic Differentiation-General Purpose Research Simulator (AD-GPRS) 40
Average absolute percent deviation (AAPD) 1, 2, 7

B

Base-neutralizing capacity (BNC) 119
Binary interaction parameters (BIPs) 6
Building Information Modelling (BIM) 178

C

Capillary pressure 100
Component Object Model (COM) 179
Compositional space adaptive tabulation (CSAT) 42
Courant–Friedrich–Lewy (CFL) 51
Cubic-plus-association (CPA) 84, 87, 109

D

Dynamic link library (DLL) 115, 117, 121

E

Empirically derived correlations or equations of state (EOSs) 2

Engelhard TitanoSilicate (ETS) 223

Enhanced oil recovery (EOR) 41

F

Finite differential method (FDM) 145

Fully Implicit Method (FIM) 43, 76

G

Gas relative 107, 110

H

High temperature 207, 208, 209, 216, 224, 225

Hydrocarbon mixtures 84, 87, 93, 109

I

IMplicit Pressure Explicit Saturation (IMPES) 43

International Energy Agency (IEA) 84

International Zeolite Association (IZA) 221

L

Lawal-Lake-Silberberg (LLS) 1, 2

Liquid methanol 104

Liquified Natural Gas (LNG) 205

M

Metal-Organic Frameworks (MOFs) 204

Miscible gas injection 26

Molecular weight (MW) 6

N

Natural Gas (NG) 203, 204

O

Object oriented programming (OOP) 117, 121

Oil in place (OIP) 40

P

Peng Robison (PR) 2

Percent deviation (PD) 7

Pressure Swing Adsorption (PSA) 203, 204

Process systems engineering (PSE) 41

R

Reinforced Concrete (RC) 178

Reservoir fluids 28

S

Single-phase 85, 100, 102, 103, 109

Soave-Redlich-Kwong (SRK) 1, 2

Society of Petroleum Engineers (SPE) 40

Solvent Thermal Resources Innovation Process (STRIP) 40

Steam Methane Reforming (SMR) 209

Structure Directing Agents (SDAs) 217

T

Temperature Swing Adsorption(TSA) 209

Trebble-Bishnoi (TB) 3
TriEthyleneGlycol (TEG) 207

V

Vapor-liquid-liquid equilibrium
    (VLLE) 58
Visual Basic for Applications
    (VBA) 179, 181

W

Water/rock chemical interaction
    (WRCI) 139
Water saturation 99, 100, 101,
    108, 110

X

XML (Extensible Mark-up Lan-
    guage) 179